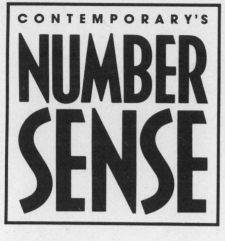

CONTEMPORARY'S

NUMBER SENSE

Discovering Basic Math Concepts

Fraction Multiplication & Division

Allan D. Suter

Project Editors
Kathy Osmus
Caren Van Slyke

CONTEMPORARY BOOKS

CHICAGO

No part of this publication may be reproduced, stored in
a retrieval system, or transmitted in any form or by any
means, without the prior written permission of the
publisher.

Published by Contemporary Books, Inc.
180 North Michigan Avenue, Chicago, Illinois 60601
Manufactured in the United States of America
International Standard Book Number: 0-8092-4224-9

Published simultaneously in Canada by
Beaverbooks, Ltd.
195 Allstate Parkway
Valleywood Business Park
Markham, Ontario L3R 4T8
Canada

Editorial Director	*Cover Design*
Caren Van Slyke	Lois Koehler
Editorial	*Illustrator*
Seija Suter	Ophelia M. Chambliss-Jones
Sarah Conroy	
Ellen Frechette	*Art & Production*
Steve Miller	Princess Louise El
Robin O'Connor	Jan Geist
Janice Bryant	
Leah Mayes	*Typography*
	J•B Typesetting
Editorial Production Manager	St. Charles, Illinois
Norma Fioretti	
Production Editor	
Craig Bolt	

Cover photo © C. C. Cain Photography.

Dedicated to our friend, Pat Reid

Contents

6 Mixed Problem Solving

7 Life-Skills Math

Fractions of a Set

1. To find $\frac{1}{3}$ of 6
   - Divide 6 into 3 equal groups
   - Shade one group

   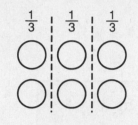

   a) Shade $\frac{1}{3}$ of 6

   b) $\frac{1}{3}$ of 6 = _2_

2.

   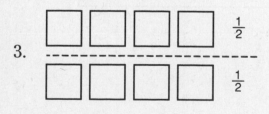

   a) Shade $\frac{1}{5}$ of 5

   b) $\frac{1}{5}$ of 5 = ____

3.

   a) Shade $\frac{1}{2}$ of 8

   b) $\frac{1}{2}$ of 8 = ____

4. To find $\frac{1}{4}$ of 12
   - Divide 12 into 4 equal groups
   - Shade one group

   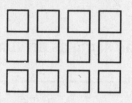

   a) Shade $\frac{1}{4}$ of 12

   b) $\frac{1}{4}$ of 12 = ____

5.

   a) Shade $\frac{1}{5}$ of 10

   b) $\frac{1}{5}$ of 10 = ____

6.

   a) Shade $\frac{1}{3}$ of 9

   b) $\frac{1}{3}$ of 9 = ____

Finding One Part of a Set

1. Shade $\frac{1}{2}$ of 8.

2. Is $\frac{1}{2}$ of 8 the same as dividing 8 by 2? _____
   yes or no

   Finding one half of any number is the same as dividing it by 2.

3. What is $\frac{1}{2}$ of $10.00? ____

4. What is $\frac{1}{2}$ of $20.00? ____

5. What is $\frac{1}{2}$ of $5.00? ____

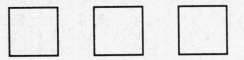

6. Shade $\frac{1}{3}$ of 3.

7. Is $\frac{1}{3}$ of 3 the same as dividing 3 by 3? _____
   yes or no

   Finding one third of any number is the same as dividing it by 3.

8. What is $\frac{1}{3}$ of $9.00? ____

9. What is $\frac{1}{3}$ of $30.00? ____

10. What is $\frac{1}{3}$ of $3.00? ____

11. What is $\frac{1}{3}$ of $21.00? ____

12. What is $\frac{1}{3}$ of $6.00? ____

Practice Helps

Find the "fraction" of a number by dividing it by the denominator.

1. $\frac{1}{6}$ of $12 = 12 \div 6 = \underline{}$
   fill in
   ↑ divide by the denominator

2. $\frac{1}{5}$ of $20 = \square \div \square = \underline{}$
   fill in

3. $\frac{1}{10}$ of $10 = \square \div \square = \underline{}$
   fill in

4. $\frac{1}{7}$ of $14 = \square \div \square = \underline{}$
   fill in

5. $\frac{1}{3}$ of $3 =$

6. $\frac{1}{8}$ of $64 =$

7. $\frac{1}{9}$ of $81 =$

8. $\frac{1}{4}$ of $40 =$

9. $\frac{1}{7}$ of $56 =$

10. $\frac{1}{10}$ of $50 =$

11. $\frac{1}{3}$ of $12 = 12 \div 3 = \underline{}$
    fill in

12. $\frac{1}{6}$ of $24 = \square \div \square = \underline{}$
    fill in

13. $\frac{1}{8}$ of $32 = \square \div \square = \underline{}$
    fill in

14. $\frac{1}{7}$ of $21 = \square \div \square = \underline{}$
    fill in

15. $\frac{1}{5}$ of $10 =$

16. $\frac{1}{10}$ of $100 =$

17. $\frac{1}{8}$ of $48 =$

18. $\frac{1}{4}$ of $8 =$

19. $\frac{1}{7}$ of $49 =$

20. $\frac{1}{5}$ of $5 =$

Shade Fractions of the Sets

1. To find $\frac{2}{3}$ of 6

   • Divide 6 into 3 equal groups
   • Shade 2 groups

   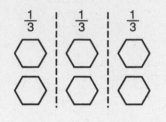

   a) Shade $\frac{2}{3}$ of 6

   b) $\frac{2}{3}$ of 6 = _4_

2.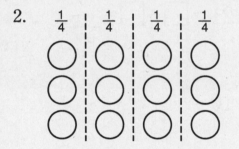

   a) Shade $\frac{3}{4}$ of 12

   b) $\frac{3}{4}$ of 12 = ____

3.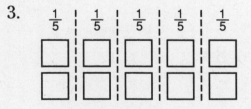

   a) Shade $\frac{4}{5}$ of 10

   b) $\frac{4}{5}$ of 10 = ____

4. To find $\frac{3}{4}$ of 8

   • Divide 8 into 4 equal groups
   • Shade 3 groups

   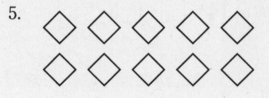

   a) Shade $\frac{3}{4}$ of 8

   b) $\frac{3}{4}$ of 8 = ____

5.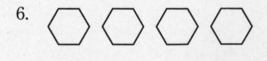

   a) Shade $\frac{3}{5}$ of 10

   b) $\frac{3}{5}$ of 10 = ____

6.

   a) Shade $\frac{3}{4}$ of 4

   b) $\frac{3}{4}$ of 4 = ____

Finding a Fraction of a Set

1.

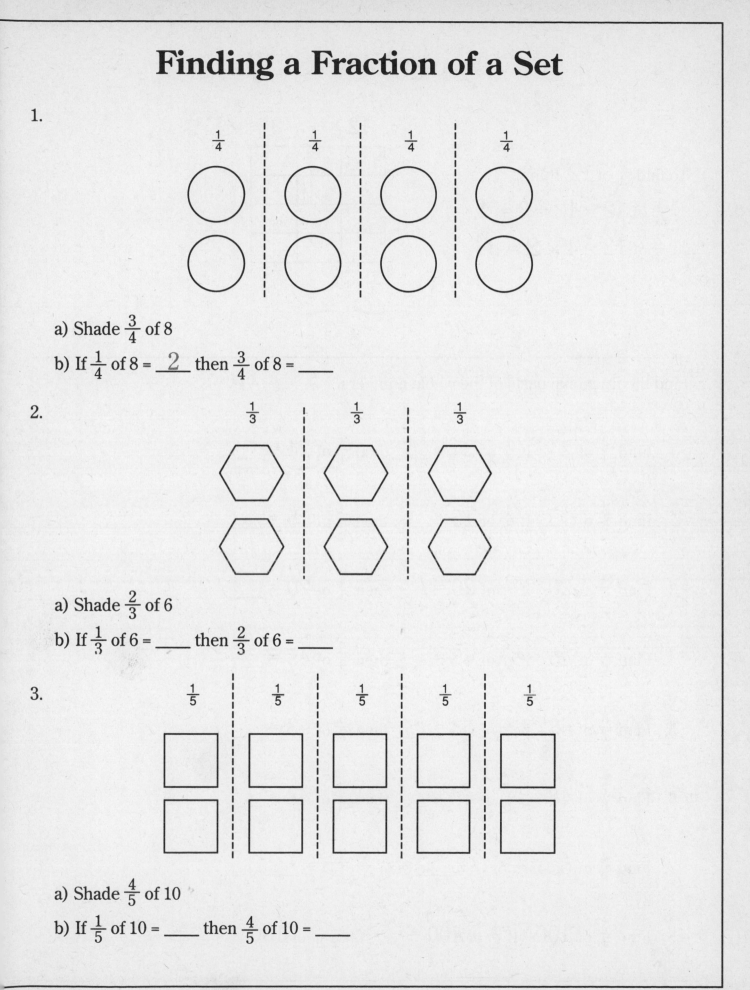

   a) Shade $\frac{3}{4}$ of 8

   b) If $\frac{1}{4}$ of 8 = __2__ then $\frac{3}{4}$ of 8 = ____

2.

   a) Shade $\frac{2}{3}$ of 6

   b) If $\frac{1}{3}$ of 6 = ____ then $\frac{2}{3}$ of 6 = ____

3.

   a) Shade $\frac{4}{5}$ of 10

   b) If $\frac{1}{5}$ of 10 = ____ then $\frac{4}{5}$ of 10 = ____

5

Apply Your Skills

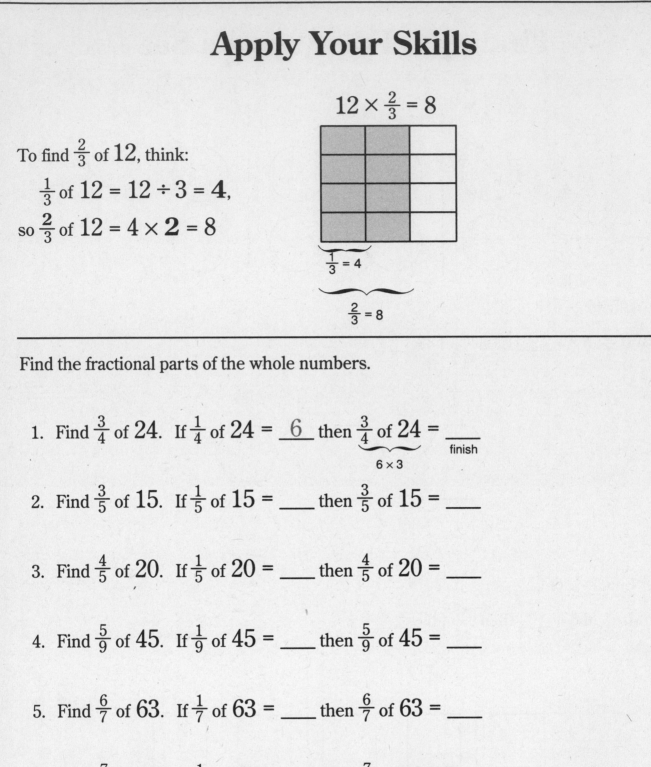

$$12 \times \frac{2}{3} = 8$$

To find $\frac{2}{3}$ of 12, think:

$\frac{1}{3}$ of 12 = 12 ÷ 3 = **4**,

so $\frac{2}{3}$ of 12 = 4 × **2** = 8

$\frac{1}{3}$ = 4

$\frac{2}{3}$ = 8

Find the fractional parts of the whole numbers.

1. Find $\frac{3}{4}$ of 24. If $\frac{1}{4}$ of 24 = __6__ then $\frac{3}{4}$ of 24 = ___
   finish
   6 × 3

2. Find $\frac{3}{5}$ of 15. If $\frac{1}{5}$ of 15 = ___ then $\frac{3}{5}$ of 15 = ___

3. Find $\frac{4}{5}$ of 20. If $\frac{1}{5}$ of 20 = ___ then $\frac{4}{5}$ of 20 = ___

4. Find $\frac{5}{9}$ of 45. If $\frac{1}{9}$ of 45 = ___ then $\frac{5}{9}$ of 45 = ___

5. Find $\frac{6}{7}$ of 63. If $\frac{1}{7}$ of 63 = ___ then $\frac{6}{7}$ of 63 = ___

6. Find $\frac{7}{8}$ of 48. If $\frac{1}{8}$ of 48 = ___ then $\frac{7}{8}$ of 48 = ___

7. Find $\frac{4}{5}$ of 5. If $\frac{1}{5}$ of 5 = ___ then $\frac{4}{5}$ of 5 = ___

8. Find $\frac{7}{10}$ of 100. If $\frac{1}{10}$ of 100 = ___ then $\frac{7}{10}$ of 100 = ___

Multiplication Is Repeated Addition

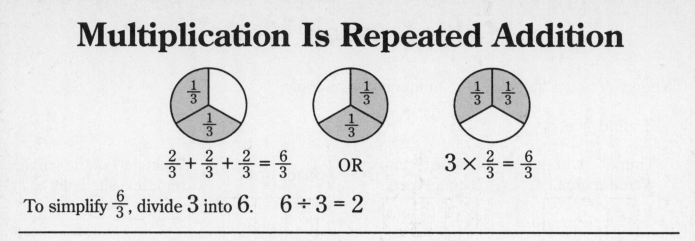

$$\frac{2}{3} + \frac{2}{3} + \frac{2}{3} = \frac{6}{3} \qquad \text{OR} \qquad 3 \times \frac{2}{3} = \frac{6}{3}$$

To simplify $\frac{6}{3}$, divide 3 into 6. $6 \div 3 = 2$

Show each example as an addition and multiplication problem.

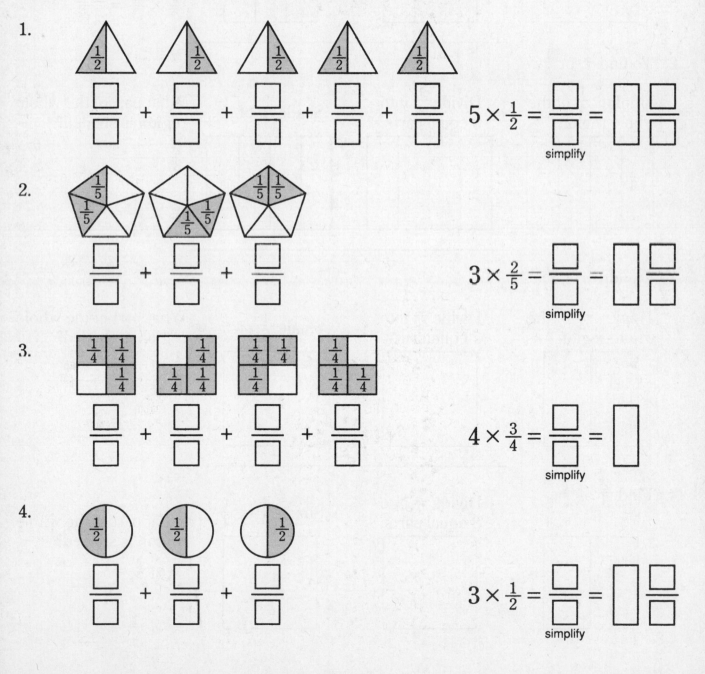

1.

$$\frac{\square}{\square} + \frac{\square}{\square} + \frac{\square}{\square} + \frac{\square}{\square} + \frac{\square}{\square} \qquad 5 \times \frac{1}{2} = \frac{\square}{\square} = \square \frac{\square}{\square}$$
simplify

2.

$$\frac{\square}{\square} + \frac{\square}{\square} + \frac{\square}{\square} \qquad 3 \times \frac{2}{5} = \frac{\square}{\square} = \square \frac{\square}{\square}$$
simplify

3.

$$\frac{\square}{\square} + \frac{\square}{\square} + \frac{\square}{\square} + \frac{\square}{\square} \qquad 4 \times \frac{3}{4} = \frac{\square}{\square} = \square$$
simplify

4.

$$\frac{\square}{\square} + \frac{\square}{\square} + \frac{\square}{\square} \qquad 3 \times \frac{1}{2} = \frac{\square}{\square} = \square \frac{\square}{\square}$$
simplify

Multiplication Models

When you find a "fraction of" a number, you multiply.

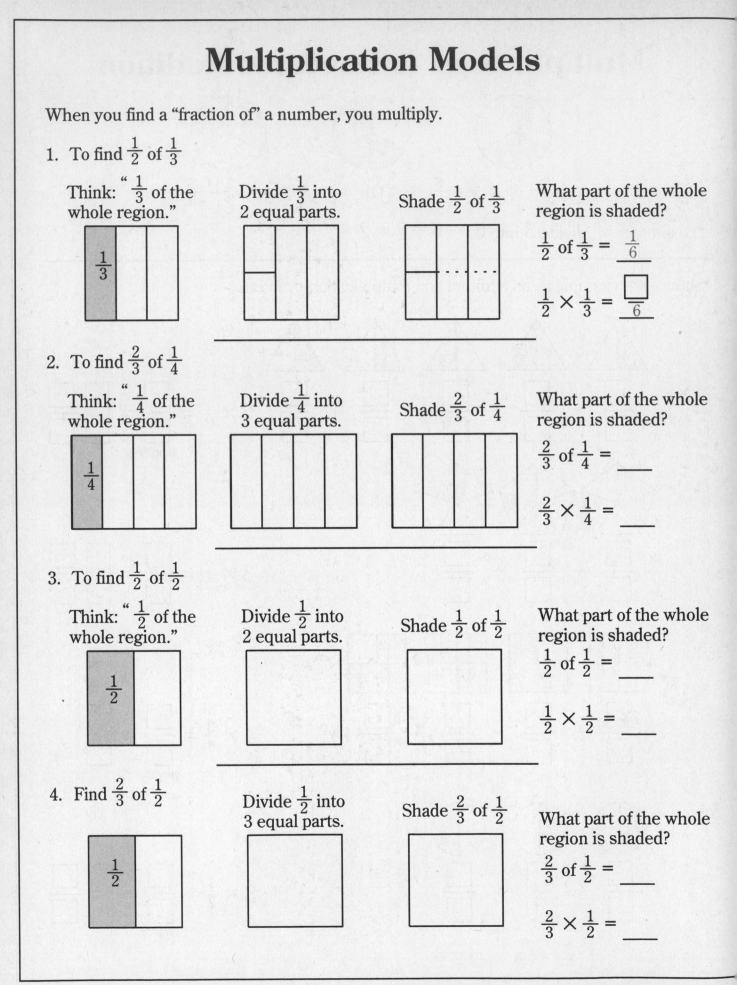

1. To find $\frac{1}{2}$ of $\frac{1}{3}$

   Think: "$\frac{1}{3}$ of the whole region."

   $\frac{1}{3}$

   Divide $\frac{1}{3}$ into 2 equal parts.

   Shade $\frac{1}{2}$ of $\frac{1}{3}$

   What part of the whole region is shaded?

   $\frac{1}{2}$ of $\frac{1}{3}$ = $\frac{1}{6}$

   $\frac{1}{2} \times \frac{1}{3}$ = $\frac{\square}{6}$

2. To find $\frac{2}{3}$ of $\frac{1}{4}$

   Think: "$\frac{1}{4}$ of the whole region."

   $\frac{1}{4}$

   Divide $\frac{1}{4}$ into 3 equal parts.

   Shade $\frac{2}{3}$ of $\frac{1}{4}$

   What part of the whole region is shaded?

   $\frac{2}{3}$ of $\frac{1}{4}$ = ____

   $\frac{2}{3} \times \frac{1}{4}$ = ____

3. To find $\frac{1}{2}$ of $\frac{1}{2}$

   Think: "$\frac{1}{2}$ of the whole region."

   $\frac{1}{2}$

   Divide $\frac{1}{2}$ into 2 equal parts.

   Shade $\frac{1}{2}$ of $\frac{1}{2}$

   What part of the whole region is shaded?

   $\frac{1}{2}$ of $\frac{1}{2}$ = ____

   $\frac{1}{2} \times \frac{1}{2}$ = ____

4. Find $\frac{2}{3}$ of $\frac{1}{2}$

   $\frac{1}{2}$

   Divide $\frac{1}{2}$ into 3 equal parts.

   Shade $\frac{2}{3}$ of $\frac{1}{2}$

   What part of the whole region is shaded?

   $\frac{2}{3}$ of $\frac{1}{2}$ = ____

   $\frac{2}{3} \times \frac{1}{2}$ = ____

Multiplying Fractions and Whole Numbers

Multiply $\frac{2}{3} \times 3$.

This means "find $\frac{2}{3}$ of 3."

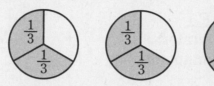

STEP 1
Write the whole number in fraction form.

$$\frac{2}{3} \times \frac{3}{1}$$

STEP 2
Multiply the numerators and the denominators.

$$\frac{2}{3} \times \frac{3}{1} = \frac{6}{3}$$

STEP 3
Simplify.

$$\frac{6}{3} = 6 \div 3 = 2$$

Multiply the numbers. Write answers in simplest form.

1. $\frac{1}{4} \times 5 = \frac{1}{4} \times \frac{5}{1} = \frac{5}{4} = \underline{1\frac{1}{4}}$
   simplify

2. $\frac{2}{5} \times 4 = \frac{2}{5} \times \frac{4}{1} =$

3. $\frac{5}{6} \times 3 =$

4. $\frac{1}{5} \times 8 =$

5. $7 \times \frac{4}{5} =$

6. $4 \times \frac{3}{4} =$

7. $\frac{2}{3} \times 8 =$

8. $6 \times \frac{1}{3} =$

9. $\frac{3}{5} \times 9 = \frac{3}{5} \times \frac{9}{1} = \frac{27}{5} = \underline{}$
   simplify

10. $6 \times \frac{2}{5} =$

11. $\frac{4}{5} \times 2 =$

12. $\frac{1}{6} \times 4 =$

13. $\frac{3}{4} \times 7 =$

14. $3 \times \frac{3}{8} =$

15. $5 \times \frac{3}{8} =$

16. $\frac{1}{2} \times 36 =$

A Fraction Times a Fraction

These examples show how to multiply fractions.

<u>EXAMPLE 1</u>
Multiply the numerators.
Multiply the denominators.

$$\frac{3}{5} \times \frac{1}{4} = \frac{3}{20}$$

<u>EXAMPLE 2</u>
Multiply. Write in simplest form.

$$\frac{3}{4} \times \frac{5}{6} = \frac{15}{24} \qquad \frac{15}{24} \div \frac{3}{3} = \frac{5}{8}$$

---

1. $\frac{1}{8} \times \frac{1}{3} = \dfrac{\boxed{1}}{\boxed{24}}$

    finish

2. $\frac{5}{6} \times \frac{1}{3} =$

3. $\frac{1}{2} \times \frac{1}{4} =$

4. $\frac{5}{6} \times \frac{1}{5} =$

5. $\frac{1}{2} \times \frac{2}{5} =$

6. $\frac{3}{4} \times \frac{1}{2} =$

7. $\frac{2}{3} \times \frac{3}{4} =$

8. $\frac{3}{5} \times \frac{2}{3} = \frac{6}{15} = \dfrac{\boxed{2}}{\boxed{5}}$

    ↑
    simplify

9. $\frac{4}{5} \times \frac{1}{2} =$

10. $\frac{1}{4} \times \frac{1}{5} =$

11. $\frac{3}{4} \times \frac{5}{6} =$

12. $\frac{5}{8} \times \frac{2}{5} =$

13. $\frac{2}{7} \times \frac{1}{3} =$

14. $\frac{2}{5} \times \frac{5}{6} =$

Simplify the Fractions

A numerator from one fraction and a denominator from the other fraction may share a common factor. It is possible to simplify before multiplying. This makes multiplying easier and does not change the answer.

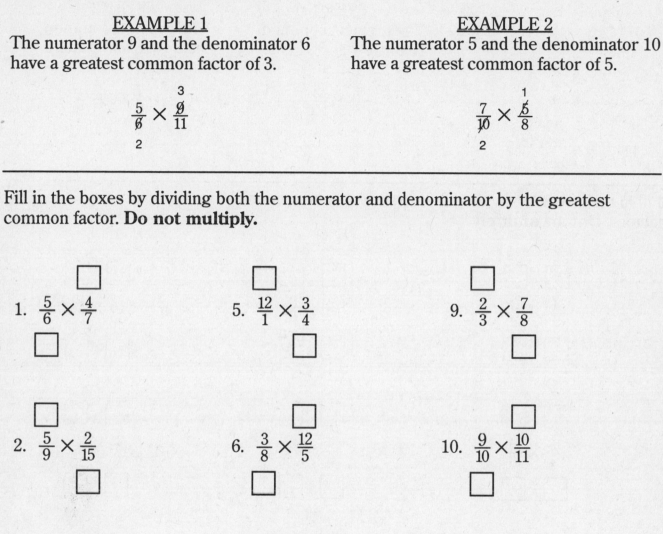

EXAMPLE 1
The numerator 9 and the denominator 6 have a greatest common factor of 3.

$$\frac{5}{\cancel{6}_2} \times \frac{\cancel{9}^3}{11}$$

EXAMPLE 2
The numerator 5 and the denominator 10 have a greatest common factor of 5.

$$\frac{7}{\cancel{10}_2} \times \frac{\cancel{5}^1}{8}$$

Fill in the boxes by dividing both the numerator and denominator by the greatest common factor. **Do not multiply.**

1. $\dfrac{\square}{\square} \quad \dfrac{5}{6} \times \dfrac{4}{7}$

2. $\dfrac{\square}{\square} \quad \dfrac{5}{9} \times \dfrac{2}{15}$

3. $\dfrac{\square}{\square} \quad \dfrac{1}{2} \times \dfrac{6}{7}$

4. $\dfrac{\square}{\square} \quad \dfrac{1}{12} \times \dfrac{3}{5}$

5. $\dfrac{\square}{\square} \quad \dfrac{12}{1} \times \dfrac{3}{4}$

6. $\dfrac{\square}{\square} \quad \dfrac{3}{8} \times \dfrac{12}{5}$

7. $\dfrac{\square}{\square} \quad \dfrac{1}{4} \times \dfrac{9}{15}$

8. $\dfrac{\square}{\square} \quad \dfrac{4}{3} \times \dfrac{17}{8}$

9. $\dfrac{\square}{\square} \quad \dfrac{2}{3} \times \dfrac{7}{8}$

10. $\dfrac{\square}{\square} \quad \dfrac{9}{10} \times \dfrac{10}{11}$

11. $\dfrac{\square}{\square} \quad \dfrac{4}{5} \times \dfrac{15}{17}$

12. $\dfrac{\square}{\square} \quad \dfrac{8}{9} \times \dfrac{1}{4}$

Simplify First

Sometimes you can simplify more than once.

$$\frac{7}{8} \times \frac{2}{21}$$

<u>STEP 1</u>
The greatest common factor of 7 and 21 is 7.

$$\frac{\overset{1}{\cancel{7}}}{8} \times \frac{2}{\underset{3}{\cancel{21}}}$$

<u>STEP 2</u>
The greatest common factor of 8 and 2 is 2.

$$\frac{\overset{1}{\cancel{7}}}{\underset{4}{\cancel{8}}} \times \frac{\overset{1}{\cancel{2}}}{\underset{3}{\cancel{21}}}$$

Fill in the boxes by dividing the numerators and denominators by the greatest common factors. **Do not multiply.**

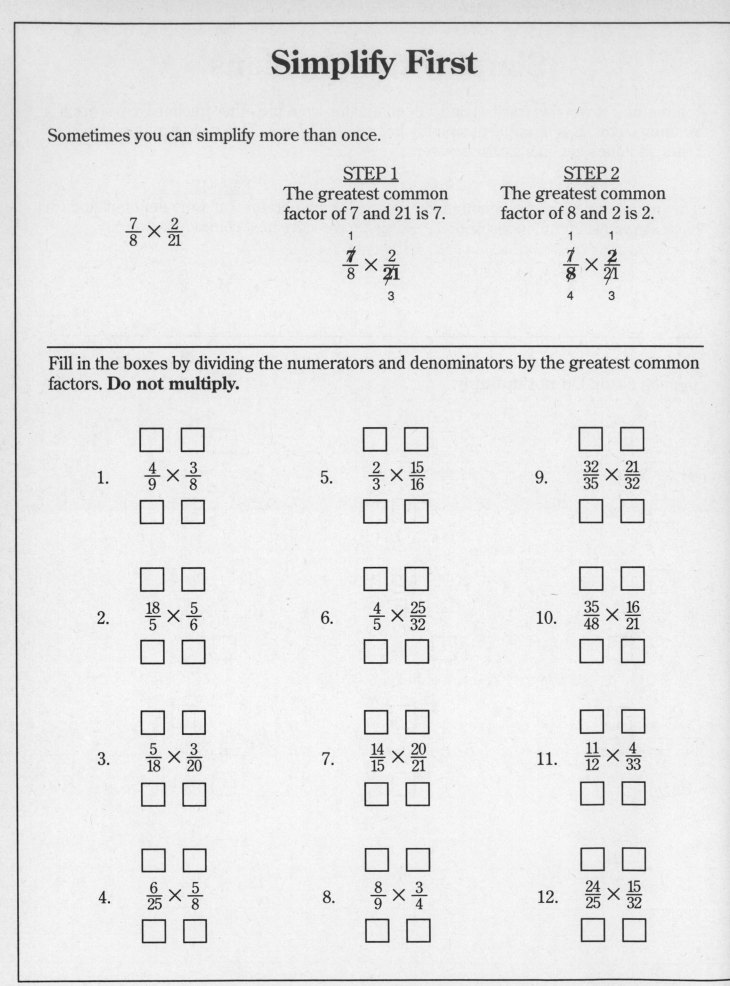

1. $\frac{4}{9} \times \frac{3}{8}$

2. $\frac{18}{5} \times \frac{5}{6}$

3. $\frac{5}{18} \times \frac{3}{20}$

4. $\frac{6}{25} \times \frac{5}{8}$

5. $\frac{2}{3} \times \frac{15}{16}$

6. $\frac{4}{5} \times \frac{25}{32}$

7. $\frac{14}{15} \times \frac{20}{21}$

8. $\frac{8}{9} \times \frac{3}{4}$

9. $\frac{32}{35} \times \frac{21}{32}$

10. $\frac{35}{48} \times \frac{16}{21}$

11. $\frac{11}{12} \times \frac{4}{33}$

12. $\frac{24}{25} \times \frac{15}{32}$

12

Simplify Before Multiplying

When a numerator and denominator share a common factor, simplify before multiplying. This makes multiplying easier and does not change the answer.

Multiply $\frac{2}{3} \times \frac{1}{6}$

<table>
<tr><td>STEP 1</td><td>STEP 2</td><td>STEP 3</td></tr>
<tr><td>2 and 6 share a common factor of 2.</td><td>Divide the numerator 2 and the denominator 6 by the common factor 2.</td><td>Multiply the numerators and then the denominators.</td></tr>
<tr><td>$\frac{2}{3} \times \frac{1}{6}$</td><td>$\frac{\overset{1}{\cancel{2}}}{3} \times \frac{1}{\underset{3}{\cancel{6}}}$</td><td>$\frac{\overset{1}{\cancel{2}}}{3} \times \frac{1}{\underset{3}{\cancel{6}}} = \frac{1}{3} \times \frac{1}{3} = \frac{1}{9}$</td></tr>
</table>

Multiply the fractions. Use common factors to simplify before you multiply.

1. $\frac{\overset{1}{\cancel{3}}}{4} \times \frac{1}{\underset{2}{\cancel{6}}} = \frac{1}{8}$

6. $\frac{\overset{1}{\cancel{6}}}{\underset{3}{\cancel{6}}} \times \frac{\overset{1}{\cancel{2}}}{\underset{1}{\cancel{3}}} =$

11. $\frac{1}{9} \times 18 = \frac{1}{\underset{1}{\cancel{9}}} \times \frac{\overset{2}{\cancel{18}}}{1} =$

2. $\frac{5}{8} \times \frac{4}{10} =$

7. $\frac{8}{9} \times \frac{3}{4} =$

12. $\frac{3}{8} \times \frac{4}{9} =$

3. $\frac{2}{3} \times \frac{6}{7} =$

8. $\frac{6}{10} \times 8 =$

13. $\frac{3}{14} \times \frac{7}{9} =$

4. $\frac{4}{7} \times \frac{1}{2} =$

9. $\frac{7}{8} \times \frac{4}{5} =$

14. $\frac{1}{6} \times \frac{3}{5} =$

5. $\frac{7}{8} \times 14 =$

10. $\frac{5}{7} \times \frac{3}{10} =$

15. $\frac{3}{5} \times \frac{5}{9} =$

Practice Your Skills

Multiply the numbers. Write the answers in simplest form.

1. $\frac{5}{7} \times \frac{1}{2} =$

2. $5 \times \frac{2}{3} =$

3. $\frac{2}{3} \times \frac{1}{6} =$

4. $\frac{1}{5} \times 7 =$

5. $\frac{3}{8} \times \frac{4}{5} =$

6. $\frac{2}{5} \times \frac{5}{6} =$

7. $\frac{1}{2} \times \frac{4}{5} =$

8. $\frac{1}{4} \times 2 =$

9. $\frac{1}{6} \times \frac{2}{3} =$

10. $\frac{2}{3} \times 9 =$

11. $\frac{5}{6} \times 3 =$

12. $\frac{2}{3} \times \frac{3}{4} =$

13. $\frac{5}{6} \times 4 =$

14. $7 \times \frac{3}{5} =$

15. $\frac{2}{5} \times \frac{3}{10} =$

16. $\frac{5}{6} \times \frac{3}{4} =$

17. $\frac{5}{8} \times 6 =$

18. $\frac{1}{2} \times \frac{1}{3} =$

19. $\frac{5}{6} \times \frac{6}{15} =$

20. $\frac{4}{9} \times \frac{3}{10} =$

Fractions Equal to Whole Numbers

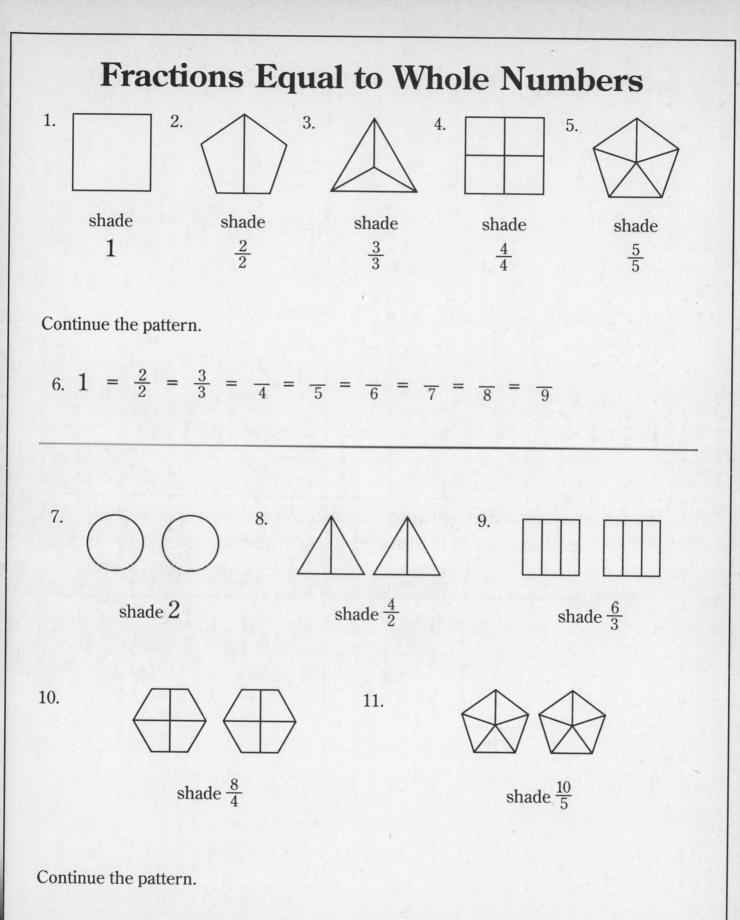

1. shade **1**

2. shade $\frac{2}{2}$

3. shade $\frac{3}{3}$

4. shade $\frac{4}{4}$

5. shade $\frac{5}{5}$

Continue the pattern.

6. $1 = \frac{2}{2} = \frac{3}{3} = \frac{}{4} = \frac{}{5} = \frac{}{6} = \frac{}{7} = \frac{}{8} = \frac{}{9}$

7. shade **2**

8. shade $\frac{4}{2}$

9. shade $\frac{6}{3}$

10. shade $\frac{8}{4}$

11. shade $\frac{10}{5}$

Continue the pattern.

12. $2 = \frac{4}{2} = \frac{6}{3} = \frac{}{4} = \frac{}{5} = \frac{}{6} = \frac{}{7} = \frac{}{8} = \frac{}{9}$

Changing Whole Numbers to Fractions

1. shade $\frac{6}{2}$ 　　　　2. shade $\frac{9}{3}$ 　　　　3. shade 3

Do you see the pattern? Continue the pattern.

(3×2)　(3×3)　(3×4)　(3×5)　(3×6)　(3×7)　(3×8)

4. $3 = \frac{6}{2} = \frac{9}{3} = \frac{}{4} = \frac{}{5} = \frac{}{6} = \frac{}{7} = \frac{}{8}$

(4×2)　(4×3)　(4×4)

5. $4 = \frac{8}{2} = \frac{}{3} = \frac{}{4} = \frac{}{5} = \frac{}{6} = \frac{}{7} = \frac{}{8}$

Change each whole number to an improper fraction. Use the given denominator as a guide.

(1×8) 　　　　　　(9×3) 　　　　　　(15×2)

6. $1 = \frac{}{8}$ 　　11. $9 = \frac{}{3}$ 　　16. $15 = \frac{}{2}$

7. $3 = \frac{}{3}$ 　　12. $4 = \frac{}{8}$ 　　17. $7 = \frac{}{5}$

8. $4 = \frac{}{10}$ 　　13. $6 = \frac{}{10}$ 　　18. $9 = \frac{}{4}$

9. $8 = \frac{}{4}$ 　　14. $7 = \frac{}{7}$ 　　19. $8 = \frac{}{3}$

10. $5 = \frac{}{6}$ 　　15. $7 = \frac{}{3}$ 　　20. $4 = \frac{}{4}$

Change Mixed Numbers
to Improper Fractions

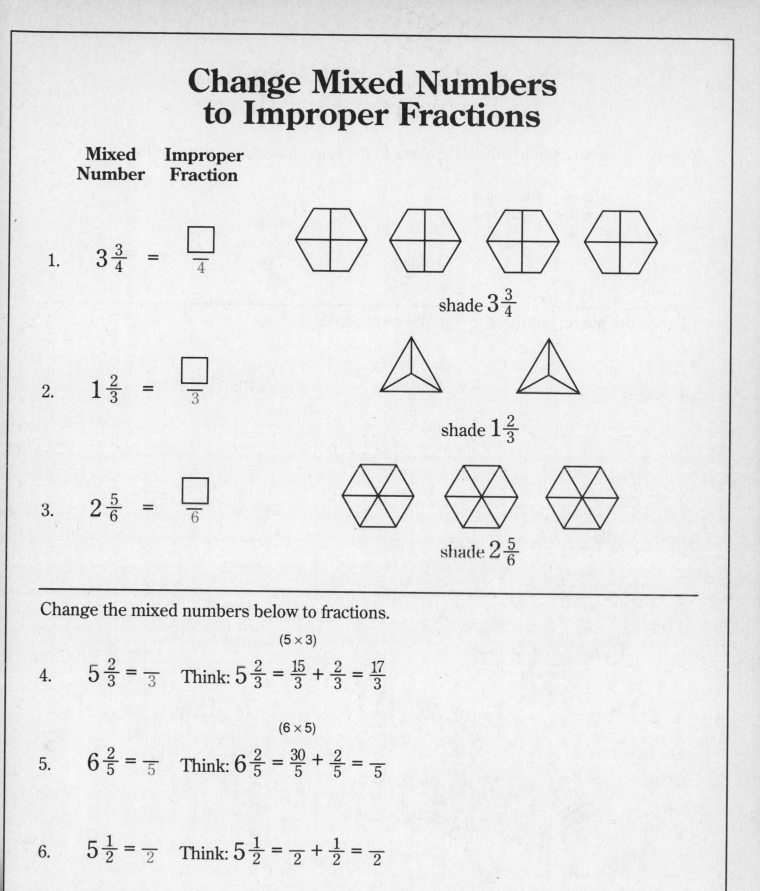

	Mixed Number	Improper Fraction

1. $3\frac{3}{4}$ = $\dfrac{\square}{4}$

shade $3\frac{3}{4}$

2. $1\frac{2}{3}$ = $\dfrac{\square}{3}$

shade $1\frac{2}{3}$

3. $2\frac{5}{6}$ = $\dfrac{\square}{6}$

shade $2\frac{5}{6}$

Change the mixed numbers below to fractions.

(5×3)

4. $5\frac{2}{3} = \dfrac{}{3}$ Think: $5\frac{2}{3} = \dfrac{15}{3} + \dfrac{2}{3} = \dfrac{17}{3}$

(6×5)

5. $6\frac{2}{5} = \dfrac{}{5}$ Think: $6\frac{2}{5} = \dfrac{30}{5} + \dfrac{2}{5} = \dfrac{}{5}$

6. $5\frac{1}{2} = \dfrac{}{2}$ Think: $5\frac{1}{2} = \dfrac{}{2} + \dfrac{1}{2} = \dfrac{}{2}$

7. $3\frac{5}{8} = \dfrac{}{8}$ Think: $3\frac{5}{8} = \dfrac{}{8} + \dfrac{5}{8} = \dfrac{}{8}$

17

More Practice Changing to Improper Fractions

You may need to change mixed numbers to improper fractions before multiplying.

$$9\frac{2}{3} = \frac{29}{3}$$

Think: $9\frac{2}{3} = \frac{27}{3} \ \overset{(9 \times 3)}{} + \frac{2}{3} = \frac{29}{3}$

Change the mixed numbers to improper fractions.

1. $2\frac{1}{2} = \frac{5}{2}$

2. $6\frac{5}{6} = \frac{41}{6}$

3. $1\frac{5}{9} = \frac{}{9}$

4. $2\frac{1}{4} = \frac{}{4}$

5. $7\frac{2}{7} = -$

6. $9\frac{2}{3} = -$

7. $9\frac{1}{3} = \frac{}{3}$

8. $2\frac{5}{8} = \frac{}{8}$

9. $5\frac{3}{4} = -$

10. $7\frac{5}{6} = -$

11. $5\frac{1}{5} = -$

12. $3\frac{5}{6} = -$

13. $3\frac{2}{3} = \frac{}{3}$

14. $12\frac{1}{2} = -$

15. $9\frac{3}{4} = -$

16. $15\frac{2}{3} = -$

17. $4\frac{1}{8} = -$

18. $8\frac{2}{5} = -$

Rename the Mixed Number

To multiply a mixed number by a fraction:

Step 1: Rename the mixed number to an improper fraction.

Step 2: Multiply the fractions.

EXAMPLE 1

$$3\frac{1}{4} \times \frac{2}{5} = \frac{13}{4} \times \frac{2}{5}$$

$$= \frac{13}{\cancel{4}_2} \times \frac{\cancel{2}^1}{5} \quad \leftarrow \text{simplify before multiplying}$$

$$= \frac{13}{10} = 1\frac{3}{10}$$

EXAMPLE 2

$$\frac{3}{10} \times 2\frac{2}{3} = \frac{3}{10} \times \frac{8}{3}$$

$$= \frac{\cancel{3}^1}{\cancel{10}_5} \times \frac{\cancel{8}^4}{\cancel{3}_1} \quad \leftarrow \text{simplify both pairs of numbers}$$

$$= \frac{4}{5}$$

Rename the mixed number, simplify, and multiply.

1. $1\frac{5}{6} \times \frac{2}{7} = \frac{11}{\cancel{6}_3} \times \frac{\cancel{2}^1}{7} =$

2. $\frac{7}{9} \times 3\frac{3}{5} =$

3. $\frac{3}{10} \times 2\frac{1}{5} =$

4. $4\frac{2}{5} \times \frac{8}{11} =$

5. $2\frac{4}{5} \times \frac{5}{7} =$

6. $\frac{2}{9} \times 2\frac{1}{4} = \frac{\cancel{2}^1}{\cancel{9}_1} \times \frac{\cancel{9}^1}{\cancel{4}_2} =$

7. $1\frac{3}{4} \times \frac{6}{7} =$

8. $5\frac{1}{5} \times \frac{5}{12} =$

9. $3\frac{2}{5} \times \frac{5}{8} =$

10. $\frac{3}{7} \times 2\frac{2}{5} =$

Multiplying Mixed Numbers

To multiply a mixed number by a mixed number:

Step 1: Rename the mixed numbers as improper fractions.

Step 2: Multiply the fractions.

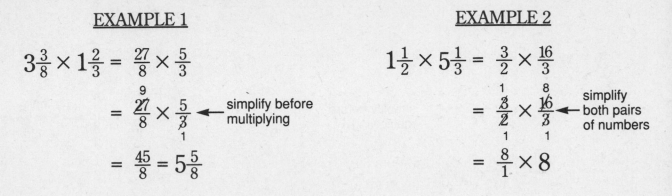

EXAMPLE 1

$$3\frac{3}{8} \times 1\frac{2}{3} = \frac{27}{8} \times \frac{5}{3}$$

$$= \frac{\overset{9}{27}}{8} \times \frac{5}{\underset{1}{3}} \quad \leftarrow \text{simplify before multiplying}$$

$$= \frac{45}{8} = 5\frac{5}{8}$$

EXAMPLE 2

$$1\frac{1}{2} \times 5\frac{1}{3} = \frac{3}{2} \times \frac{16}{3}$$

$$= \frac{\overset{1}{\cancel{3}}}{\underset{1}{2}} \times \frac{\overset{8}{\cancel{16}}}{\underset{1}{\cancel{3}}} \quad \leftarrow \text{simplify both pairs of numbers}$$

$$= \frac{8}{1} \times 8$$

Multiply the mixed numbers. Simplify before multiplying.

1. $2\frac{5}{7} \times 4\frac{2}{3} =$

2. $1\frac{1}{8} \times 3\frac{2}{9} =$

3. $1\frac{3}{4} \times 4\frac{4}{7} =$

4. $2\frac{5}{9} \times 3\frac{3}{5} =$

5. $5\frac{1}{3} \times 4\frac{1}{2} =$

6. $4\frac{5}{8} \times 1\frac{3}{5} =$

7. $1\frac{6}{7} \times 3\frac{8}{9} =$

8. $2\frac{2}{7} \times 2\frac{7}{8} =$

Master the Skills

Multiply the numbers. Write the answers in the simplest form.

1. $\frac{1}{5} \times 7 =$

2. $\frac{2}{3} \times \frac{9}{10} =$

3. $3\frac{3}{4} \times \frac{2}{3} =$

4. $\frac{5}{6} \times 12 =$

5. $\frac{2}{5} \times \frac{1}{3} =$

6. $\frac{3}{5} \times \frac{5}{12} =$

7. $1\frac{2}{5} \times \frac{3}{7} =$

8. $5\frac{3}{4} \times \frac{1}{3} =$

9. $1\frac{1}{4} \times \frac{6}{7} =$

10. $1\frac{5}{6} \times \frac{2}{7} =$

11. $1\frac{7}{8} \times 3\frac{1}{3} =$

12. $8 \times 1\frac{3}{4} =$

13. $1\frac{3}{5} \times \frac{5}{12} =$

14. $\frac{1}{2} \times 3\frac{1}{4} =$

15. $4\frac{1}{6} \times 1\frac{3}{5} =$

16. $\frac{3}{4} \times 7 =$

Multiplication Review

1. What is $\frac{1}{3}$ of $15.00?

2. $\frac{1}{5}$ of 25 =

3. a) Shade $\frac{3}{4}$ of 8

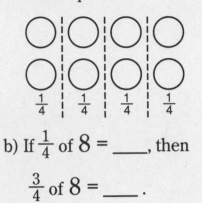

   $\frac{1}{4}$ $\frac{1}{4}$ $\frac{1}{4}$ $\frac{1}{4}$

   b) If $\frac{1}{4}$ of 8 = ____, then

   $\frac{3}{4}$ of 8 = ____ .

4. Find $\frac{3}{4}$ of 36. If $\frac{1}{4}$ of 36 = ____ ,

   then $\frac{3}{4}$ of 36 = ____ .

5.

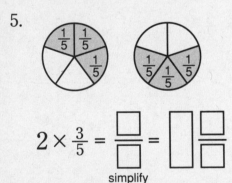

   $2 \times \frac{3}{5} = \dfrac{\square}{\square} = \square \; \dfrac{\square}{\square}$

   simplify

6. $\frac{2}{3}$ of $\frac{1}{3} = \frac{2}{3}$ ___ $\frac{1}{3}$
   symbol

7. $4 \times \frac{3}{5} =$

8. $\frac{7}{8} \times \frac{1}{3} =$

Simplify before multiplying.

9. $\frac{1}{5} \times \frac{15}{17} =$

10. $\frac{8}{9} \times \frac{3}{4} =$

11. $\frac{5}{18} \times \frac{2}{5} =$

12. $\frac{2}{3} \times 6 =$

Rename each number as an improper fraction.

13. $2 = \dfrac{}{6}$

14. $7 = \dfrac{}{4}$

15. $4\frac{3}{5} = \dfrac{}{5}$

16. $6\frac{1}{8} = \dfrac{}{8}$

Rename the mixed number, simplify, and multiply.

17. $2\frac{2}{5} \times \frac{5}{6} =$

18. $5\frac{1}{3} \times 5\frac{1}{4} =$

Find a Fraction of an Amount

When you find a **fraction** of an amount, you multiply the fraction times the amount.

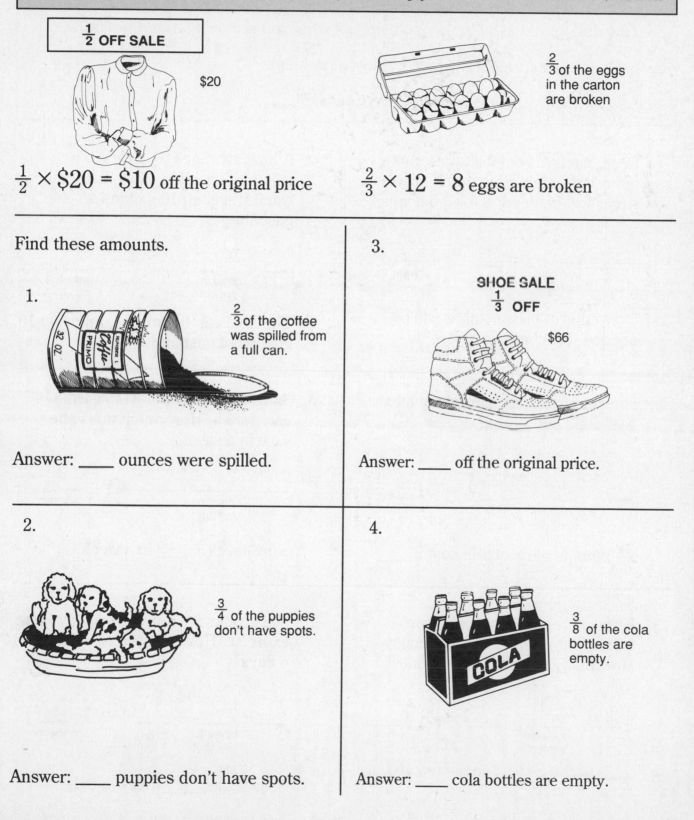

$\frac{1}{2}$ OFF SALE

$20

$\frac{2}{3}$ of the eggs in the carton are broken

$\frac{1}{2} \times \$20 = \$10$ off the original price

$\frac{2}{3} \times 12 = 8$ eggs are broken

Find these amounts.

1.

$\frac{2}{3}$ of the coffee was spilled from a full can.

Answer: _____ ounces were spilled.

3.

SHOE SALE $\frac{1}{3}$ OFF

$66

Answer: _____ off the original price.

2.

$\frac{3}{4}$ of the puppies don't have spots.

Answer: _____ puppies don't have spots.

4.

$\frac{3}{8}$ of the cola bottles are empty.

Answer: _____ cola bottles are empty.

Does the Answer Make Sense?

1. Read over the problem several times to make sure you understand it.

2. Think about the facts in the problem and what you are being asked to find.

3. Complete the number sentence for each problem.

4. Ask yourself, "Does the answer make sense?"

1. Each batch of cookies takes $\frac{1}{2}$ cup of brown sugar. How much brown sugar will be needed for 5 batches?

____ ____ ____ = ____
   operation   answer
    symbol

____ cups of brown sugar will be needed for 5 batches.

2. How much will $2\frac{2}{3}$ pounds of meat cost at $3 per pound?

____ ____ ____ = ____
   operation   answer
    symbol

$2\frac{2}{3}$ pounds of meat will cost $ ____.

3. Kitty surveyed 24 students and found that $\frac{3}{4}$ of them save money. How many students save money?

____ ____ ____ = ____
   operation   answer
    symbol

____ of the 24 students surveyed saved money.

4. A boat averages $6\frac{1}{2}$ miles to a gallon of gasoline. How many miles can it travel on 10 gallons of gasoline?

____ ____ ____ = ____
   operation   answer
    symbol

The boat can travel ____ miles on 10 gallons of gasoline.

5. Sally saves $\frac{1}{5}$ of her $145 earnings each week. How much does she save in a week?

____ ____ ____ = ____
   operation   answer
    symbol

Sally saves $ ____ in a week.

6. Seija walks $3\frac{3}{4}$ miles every day for exercise. How far does she walk in 6 days?

____ ____ ____ = ____
   operation   answer
    symbol

Seija walks ____ miles in 6 days.

Number Sentences

Write number sentences and solve the problems below.

1. A box of candy weighs $\frac{1}{2}$ pound. How much will 4 boxes weigh?

    ____ ____ ____ = ____
    <small>operation answer</small>
    <small>symbol</small>

    4 boxes of candy weigh ____ pounds.

2. Bill lost an average of $\frac{3}{4}$ of a pound every week. How many pounds did he lose in 8 weeks?

    ____ ____ ____ = ____
    <small>operation answer</small>
    <small>symbol</small>

    Bill lost ____ pounds in 8 weeks.

3. Each book measures $\frac{3}{4}$ inches thick. If 4 books are placed on top of one another, how high will the stack be?

    ____ ____ ____ = ____
    <small>operation answer</small>
    <small>symbol</small>

    The stack of books will be ____ inches high.

4. Leon's car holds $16\frac{1}{2}$ gallons of gas. If the tank is $\frac{1}{3}$ full, how much gas is in his tank?

    ____ ____ ____ = ____
    <small>operation answer</small>
    <small>symbol</small>

    There are ____ gallons of gas in the tank.

5. Each bag weighs $\frac{5}{8}$ of a pound. How much will 48 bags weigh?

    ____ ____ ____ = ____
    <small>operation answer</small>
    <small>symbol</small>

    48 bags will weigh ____ pounds.

6. Jill drinks $\frac{3}{4}$ of a glass of orange juice each morning for breakfast. How much does she drink in 28 days?

    ____ ____ ____ = ____
    <small>operation answer</small>
    <small>symbol</small>

    Jill drinks ____ glasses of orange juice in 28 days.

7. Sue bought a $24 sweater that was marked $\frac{1}{3}$ off. How much was the discount?

    ____ ____ ____ = ____
    <small>operation answer</small>
    <small>symbol</small>

    The discount was $ ____.

8. Alice works $6\frac{1}{2}$ hours each day, 6 days a week. How many hours does she work each week?

    ____ ____ ____ = ____
    <small>operation answer</small>
    <small>symbol</small>

    Alice works ____ hours a week.

Divide Whole Numbers by Fractions

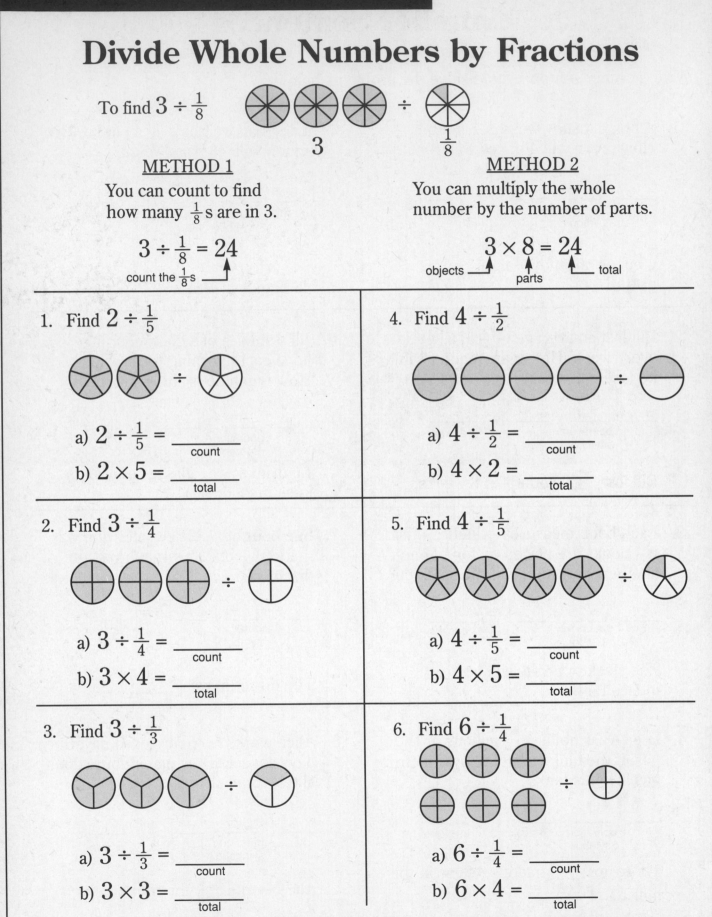

To find $3 \div \frac{1}{8}$

3 $\frac{1}{8}$

METHOD 1
You can count to find how many $\frac{1}{8}$s are in 3.

$3 \div \frac{1}{8} = 24$

count the $\frac{1}{8}$s

METHOD 2
You can multiply the whole number by the number of parts.

$3 \times 8 = 24$

objects —— parts —— total

1. Find $2 \div \frac{1}{5}$

a) $2 \div \frac{1}{5} =$ _____ count

b) $2 \times 5 =$ _____ total

4. Find $4 \div \frac{1}{2}$

a) $4 \div \frac{1}{2} =$ _____ count

b) $4 \times 2 =$ _____ total

2. Find $3 \div \frac{1}{4}$

a) $3 \div \frac{1}{4} =$ _____ count

b) $3 \times 4 =$ _____ total

5. Find $4 \div \frac{1}{5}$

a) $4 \div \frac{1}{5} =$ _____ count

b) $4 \times 5 =$ _____ total

3. Find $3 \div \frac{1}{3}$

a) $3 \div \frac{1}{3} =$ _____ count

b) $3 \times 3 =$ _____ total

6. Find $6 \div \frac{1}{4}$

a) $6 \div \frac{1}{4} =$ _____ count

b) $6 \times 4 =$ _____ total

Think About Fraction Division

Complete.

1. $2 \div \frac{1}{4} =$ __8__

2. $2 \times 4 =$ ___

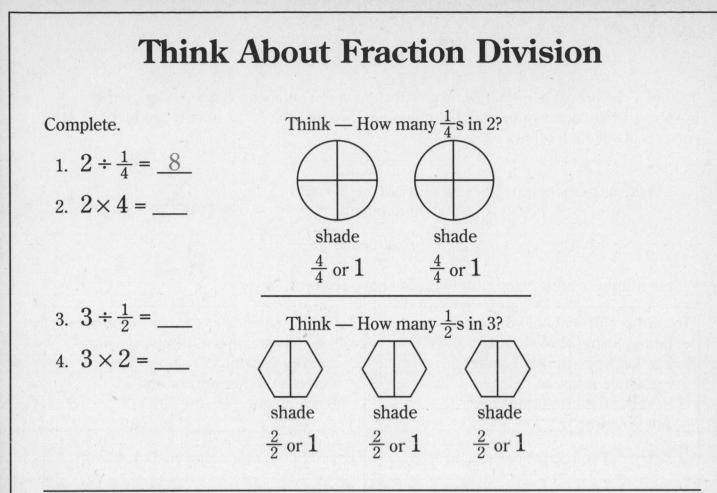

Think — How many $\frac{1}{4}$s in 2?

shade   shade

$\frac{4}{4}$ or 1 $\frac{4}{4}$ or 1

3. $3 \div \frac{1}{2} =$ ___

4. $3 \times 2 =$ ___

Think — How many $\frac{1}{2}$s in 3?

shade   shade   shade

$\frac{2}{2}$ or 1 $\frac{2}{2}$ or 1 $\frac{2}{2}$ or 1

Divide the whole number by the fraction. To do so, multiply the whole number by the number of parts.

5. $1 \div \frac{1}{4} =$ ___
Think: 1×4

9. $5 \div \frac{1}{2} =$ ___
Think: 5×2

13. $3 \div \frac{1}{5} =$ ___
Think: 3×5

6. $2 \div \frac{1}{2} =$ ___

10. $4 \div \frac{1}{5} =$ ___

14. $6 \div \frac{1}{2} =$ ___

7. $2 \div \frac{1}{3} =$ ___

11. $2 \div \frac{1}{4} =$ ___

15. $4 \div \frac{1}{4} =$ ___

8. $3 \div \frac{1}{4} =$ ___

12. $3 \div \frac{1}{3} =$ ___

16. $4 \div \frac{1}{2} =$ ___

Reciprocals

To find a reciprocal of a fraction, you "turn it over." This means you exchange the numbers in the numerator and denominator. Two numbers that have a product of 1 are reciprocals of each other.

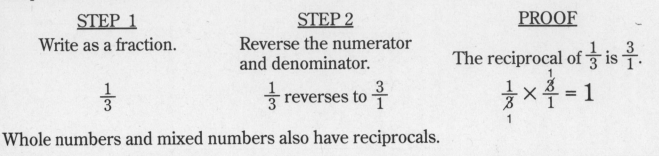

STEP 1	STEP 2	PROOF
Write as a fraction.	Reverse the numerator and denominator.	The reciprocal of $\frac{1}{3}$ is $\frac{3}{1}$.
$\frac{1}{3}$	$\frac{1}{3}$ reverses to $\frac{3}{1}$	$\frac{1}{3} \times \frac{3}{1} = 1$

Whole numbers and mixed numbers also have reciprocals.

To find reciprocals of whole numbers:	To find reciprocals of mixed numbers:
• Rename any whole number as a fraction by writing a 1 under the whole number. • Reverse the numerator and denominator.	• Rename the mixed number as an improper fraction. • Reverse the numerator and denominator.
$5 = \frac{5}{1}$ The reciprocal of $\frac{5}{1}$ is $\frac{1}{5}$ $6 = \frac{6}{1}$ The reciprocal of $\frac{6}{1}$ is $\frac{1}{6}$	$2\frac{2}{3} = \frac{8}{3}$ The reciprocal of $\frac{8}{3}$ is $\frac{3}{8}$ $5\frac{3}{4} = \frac{23}{4}$ The reciprocal of $\frac{23}{4}$ is $\frac{4}{23}$

Write the reciprocal of each number.

Number	Reciprocal		Number	Reciprocal		Number	Reciprocal
1. $\frac{3}{5}$	$\frac{5}{3}$	6. $5 = \frac{5}{1}$	$\frac{1}{5}$	11. $3\frac{1}{2} = \frac{7}{2}$	$\frac{2}{7}$
2. $\frac{3}{8}$	$\frac{\Box}{3}$	7. $7 = \frac{7}{1}$	$\frac{\Box}{7}$	12. $2\frac{3}{4} = \frac{11}{4}$	$\frac{\Box}{11}$
3. $\frac{8}{2}$	$\frac{\Box}{\Box}$	8. $3 = \frac{\Box}{\Box}$	$\frac{\Box}{\Box}$	13. $4\frac{1}{5} = \frac{\Box}{\Box}$	$\frac{\Box}{\Box}$
4. $\frac{3}{4}$	$\frac{\Box}{\Box}$	9. $4 = \frac{\Box}{\Box}$	$\frac{\Box}{\Box}$	14. $3\frac{5}{6} = \frac{\Box}{\Box}$	$\frac{\Box}{\Box}$
5. $\frac{5}{3}$	$\frac{\Box}{\Box}$	10. $15 = \frac{\Box}{\Box}$	$\frac{\Box}{\Box}$	15. $1\frac{3}{5} = \frac{\Box}{\Box}$	$\frac{\Box}{\Box}$

Dividing with Fractions

Use the drawings to answer the questions.

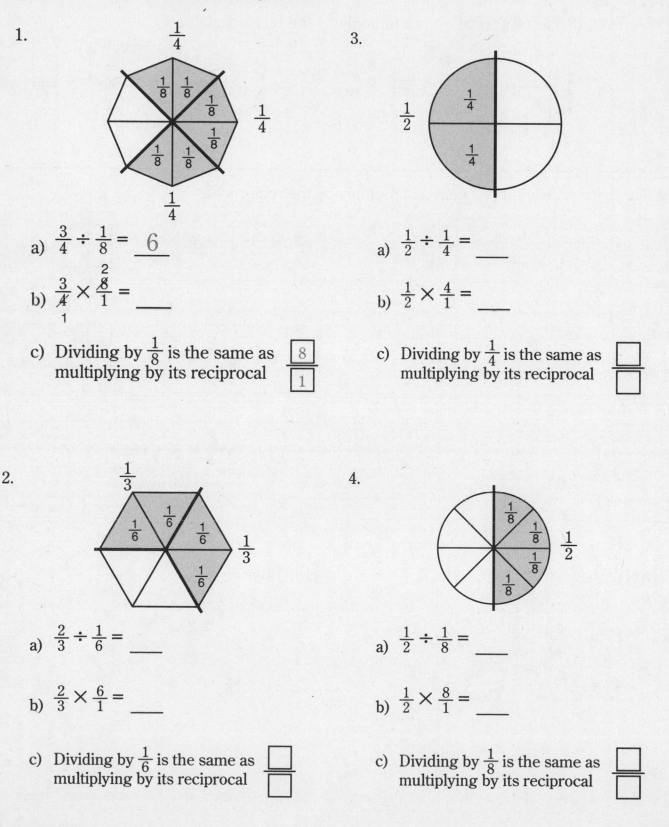

1.

a) $\dfrac{3}{4} \div \dfrac{1}{8} = \underline{\quad 6 \quad}$

b) $\dfrac{3}{\cancel{4}}^{1} \times \dfrac{\cancel{8}^{2}}{1} = \underline{\quad}$

c) Dividing by $\dfrac{1}{8}$ is the same as multiplying by its reciprocal $\dfrac{8}{1}$

2.

a) $\dfrac{2}{3} \div \dfrac{1}{6} = \underline{\quad}$

b) $\dfrac{2}{3} \times \dfrac{6}{1} = \underline{\quad}$

c) Dividing by $\dfrac{1}{6}$ is the same as multiplying by its reciprocal $\dfrac{\square}{\square}$

3.

a) $\dfrac{1}{2} \div \dfrac{1}{4} = \underline{\quad}$

b) $\dfrac{1}{2} \times \dfrac{4}{1} = \underline{\quad}$

c) Dividing by $\dfrac{1}{4}$ is the same as multiplying by its reciprocal $\dfrac{\square}{\square}$

4.

a) $\dfrac{1}{2} \div \dfrac{1}{8} = \underline{\quad}$

b) $\dfrac{1}{2} \times \dfrac{8}{1} = \underline{\quad}$

c) Dividing by $\dfrac{1}{8}$ is the same as multiplying by its reciprocal $\dfrac{\square}{\square}$

Multiply by the Reciprocal

When you divide by a fraction, you multiply by the reciprocal.

A. $\dfrac{2}{3} \div \dfrac{3}{4} = \dfrac{2}{3} \times \dfrac{4}{3} = \dfrac{\square}{\square}$

B. $\dfrac{2}{5} \div \dfrac{7}{8} = \dfrac{2}{5} \times \dfrac{8}{7} = \dfrac{\square}{\square}$

Rewrite each division problem and multiply by the reciprocal.

1. $\dfrac{1}{5} \div \dfrac{1}{4} = \dfrac{1}{5} \times \dfrac{4}{1} = \dfrac{\square}{\square}$

7. $\dfrac{1}{5} \div \dfrac{7}{8} = \dfrac{1}{5} \times \dfrac{8}{7} = \dfrac{\square}{\square}$

2. $\dfrac{1}{4} \div \dfrac{2}{3} = \dfrac{1}{4} \times \dfrac{3}{2} = \dfrac{\square}{\square}$

8. $\dfrac{3}{5} \div \dfrac{7}{8} = \dfrac{3}{5} \times \dfrac{8}{7} = \dfrac{\square}{\square}$

3. $\dfrac{2}{3} \div \dfrac{7}{8} =$

9. $\dfrac{1}{2} \div \dfrac{3}{5} =$

4. $\dfrac{1}{8} \div \dfrac{2}{3} =$

10. $\dfrac{2}{5} \div \dfrac{3}{4} =$

5. $\dfrac{4}{5} \div \dfrac{7}{8} =$

11. $\dfrac{3}{4} \div \dfrac{4}{5} =$

6. $\dfrac{1}{5} \div \dfrac{1}{3} =$

12. $\dfrac{1}{6} \div \dfrac{1}{5} =$

A Fraction Divided by a Fraction

To divide by a fraction, multiply by the reciprocal. Simplify before you multiply when necessary.

EXAMPLE	STEP 1	STEP 2	STEP 3
	Multiply by the reciprocal.	Simplify and multiply.	Write the answer in the simplest form.
$\frac{2}{3} \div \frac{2}{5}$	$\frac{2}{3} \times \frac{5}{2}$	$\frac{\overset{1}{\cancel{2}}}{3} \times \frac{5}{\underset{1}{\cancel{2}}} = \frac{5}{3}$	$\frac{5}{3} = 1\frac{2}{3}$

Follow the steps above to solve each problem.

1. $\frac{1}{3} \div \frac{5}{3} = \frac{1}{3} \times \frac{3}{5} =$

2. $\frac{5}{12} \div \frac{1}{3} =$

3. $\frac{1}{8} \div \frac{1}{3} =$

4. $\frac{2}{3} \div \frac{1}{2} =$

5. $\frac{1}{5} \div \frac{1}{4} =$

6. $\frac{9}{10} \div \frac{3}{8} =$

7. $\frac{3}{10} \div \frac{2}{5} = \frac{3}{10} \times \frac{5}{2} =$

8. $\frac{5}{4} \div \frac{2}{3} =$

9. $\frac{1}{4} \div \frac{1}{2} =$

10. $\frac{2}{3} \div \frac{1}{5} =$

11. $\frac{4}{5} \div \frac{3}{4} =$

12. $\frac{1}{2} \div \frac{1}{6} =$

Dividing by a Fraction

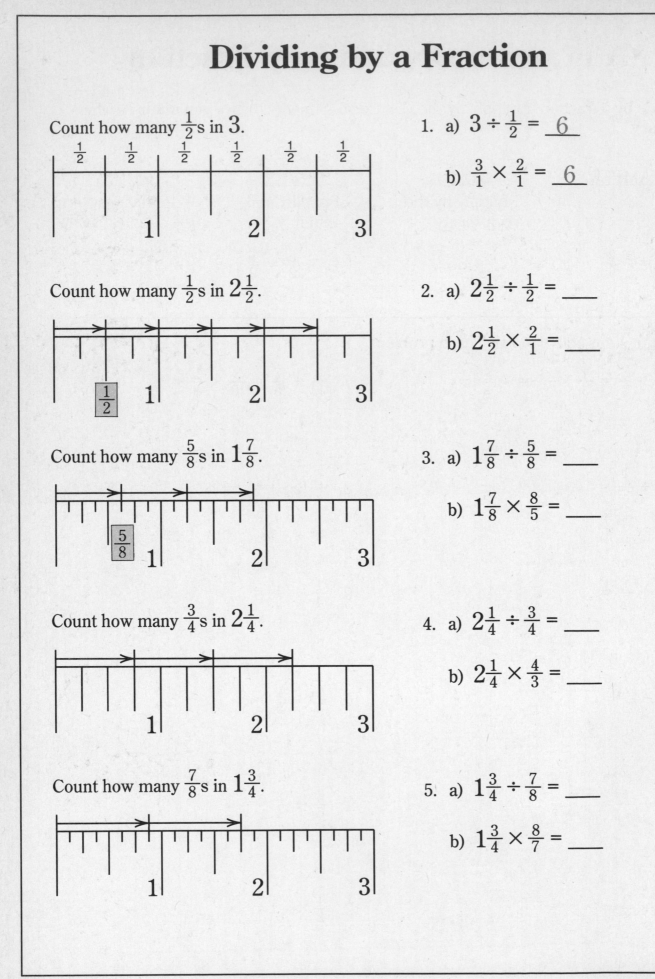

Count how many $\frac{1}{2}$s in 3.

1. a) $3 \div \frac{1}{2} =$ _6_

   b) $\frac{3}{1} \times \frac{2}{1} =$ _6_

Count how many $\frac{1}{2}$s in $2\frac{1}{2}$.

2. a) $2\frac{1}{2} \div \frac{1}{2} =$ ___

   b) $2\frac{1}{2} \times \frac{2}{1} =$ ___

Count how many $\frac{5}{8}$s in $1\frac{7}{8}$.

3. a) $1\frac{7}{8} \div \frac{5}{8} =$ ___

   b) $1\frac{7}{8} \times \frac{8}{5} =$ ___

Count how many $\frac{3}{4}$s in $2\frac{1}{4}$.

4. a) $2\frac{1}{4} \div \frac{3}{4} =$ ___

   b) $2\frac{1}{4} \times \frac{4}{3} =$ ___

Count how many $\frac{7}{8}$s in $1\frac{3}{4}$.

5. a) $1\frac{3}{4} \div \frac{7}{8} =$ ___

   b) $1\frac{3}{4} \times \frac{8}{7} =$ ___

Think About Dividing by Fractions

Use the pictures to answer the questions.

Count how many $\frac{1}{2}$s in $1\frac{1}{2}$.

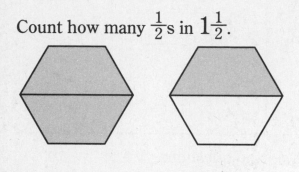

1. a) $1\frac{1}{2} \div \frac{1}{2} =$ _____

   b) $1\frac{1}{2} \times \frac{2}{1} =$ _____

Count how many $\frac{1}{3}$s in $1\frac{2}{3}$.

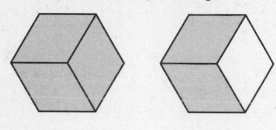

2. a) $1\frac{2}{3} \div \frac{1}{3} =$ _____

   b) $1\frac{2}{3} \times \frac{3}{1} =$ _____

Count how many $\frac{1}{6}$s in $1\frac{1}{6}$.

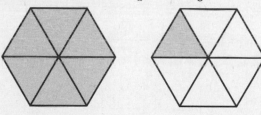

3. a) $1\frac{1}{6} \div \frac{1}{6} =$ _____

   b) $1\frac{1}{6} \times \frac{6}{1} =$ _____

Count how many $\frac{1}{6}$s in $1\frac{1}{2}$.

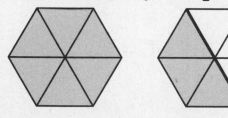

4. a) $1\frac{1}{2} \div \frac{1}{6} =$ _____

   b) $1\frac{1}{2} \times \frac{6}{1} =$ _____

Count how many $\frac{2}{3}$s in $1\frac{1}{3}$.

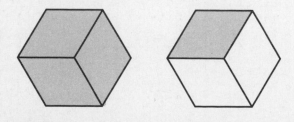

5. a) $1\frac{1}{3} \div \frac{2}{3} =$ _____

   b) $1\frac{1}{3} \times \frac{3}{2} =$ _____

Using Drawings

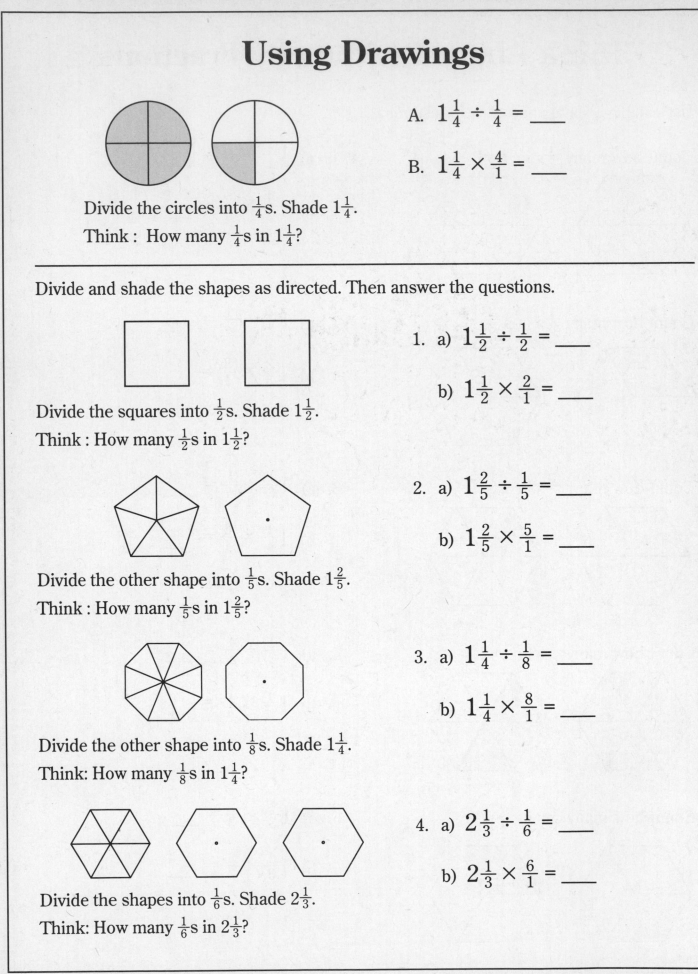

Divide the circles into $\frac{1}{4}$s. Shade $1\frac{1}{4}$.

Think : How many $\frac{1}{4}$s in $1\frac{1}{4}$?

A. $1\frac{1}{4} \div \frac{1}{4} =$ ____

B. $1\frac{1}{4} \times \frac{4}{1} =$ ____

Divide and shade the shapes as directed. Then answer the questions.

Divide the squares into $\frac{1}{2}$s. Shade $1\frac{1}{2}$.

Think : How many $\frac{1}{2}$s in $1\frac{1}{2}$?

1. a) $1\frac{1}{2} \div \frac{1}{2} =$ ____

    b) $1\frac{1}{2} \times \frac{2}{1} =$ ____

Divide the other shape into $\frac{1}{5}$s. Shade $1\frac{2}{5}$.

Think : How many $\frac{1}{5}$s in $1\frac{2}{5}$?

2. a) $1\frac{2}{5} \div \frac{1}{5} =$ ____

    b) $1\frac{2}{5} \times \frac{5}{1} =$ ____

Divide the other shape into $\frac{1}{8}$s. Shade $1\frac{1}{4}$.

Think: How many $\frac{1}{8}$s in $1\frac{1}{4}$?

3. a) $1\frac{1}{4} \div \frac{1}{8} =$ ____

    b) $1\frac{1}{4} \times \frac{8}{1} =$ ____

Divide the shapes into $\frac{1}{6}$s. Shade $2\frac{1}{3}$.

Think: How many $\frac{1}{6}$s in $2\frac{1}{3}$?

4. a) $2\frac{1}{3} \div \frac{1}{6} =$ ____

    b) $2\frac{1}{3} \times \frac{6}{1} =$ ____

Dividing a Mixed Number by a Fraction

EXAMPLE	STEP 1	STEP 2
$1\frac{1}{2} \div \frac{1}{4}$	Change the mixed number to an improper fraction.	Multiply by the reciprocal.

$$1\frac{1}{2} \div \frac{1}{4} = \frac{3}{2} \div \frac{1}{4}$$

$$\frac{3}{\underset{1}{2}} \times \frac{\overset{2}{4}}{1} = \frac{6}{1} = 6$$

Divide the mixed number by the fraction.

1. $1\frac{1}{4} \div \frac{1}{3} = \frac{5}{4} \times \frac{3}{1} =$ (reciprocal / multiply)

7. $5\frac{1}{3} \div \frac{2}{5} = \frac{16}{3} \times \frac{5}{2} =$ (reciprocal / multiply)

2. $2\frac{2}{3} \div \frac{3}{4} =$

8. $4\frac{1}{5} \div \frac{3}{4} =$

3. $4\frac{1}{2} \div \frac{3}{5} =$

9. $4\frac{3}{8} \div \frac{7}{8} =$

4. $3\frac{1}{2} \div \frac{2}{3} =$

10. $1\frac{1}{9} \div \frac{5}{9} =$

5. $1\frac{4}{5} \div \frac{3}{4} =$

11. $8\frac{1}{3} \div \frac{1}{3} =$

6. $3\frac{3}{4} \div \frac{1}{8} =$

12. $2\frac{1}{8} \div \frac{1}{6} =$

Mixed Practice

Divide by the fractions by using the reciprocals.

1. $2\frac{2}{3} \div \frac{1}{6} = \frac{8}{\cancel{3}_1} \times \frac{\cancel{6}^2}{1} =$

2. $\frac{3}{4} \div \frac{7}{8} =$

3. $2\frac{1}{3} \div \frac{1}{6} =$

4. $\frac{2}{9} \div \frac{2}{3} =$

5. $10\frac{1}{2} \div \frac{3}{5} =$

6. $\frac{3}{4} \div \frac{1}{8} =$

7. $8\frac{1}{3} \div \frac{1}{3} =$

8. $4\frac{1}{5} \div \frac{3}{5} =$

9. $3\frac{3}{4} \div \frac{3}{8} = \frac{15}{4} \times \frac{8}{3} =$

10. $\frac{4}{5} \div \frac{8}{15} =$

11. $\frac{2}{3} \div \frac{1}{6} =$

12. $6\frac{1}{4} \div \frac{5}{8} =$

13. $3\frac{1}{5} \div \frac{4}{5} =$

14. $\frac{4}{1} \div \frac{3}{7} =$

15. $2\frac{1}{4} \div \frac{3}{4} =$

16. $\frac{3}{4} \div \frac{1}{2} =$

Divide Whole Numbers by Fractions

Any whole number can be written as a fraction. Just place the whole number over a denominator of 1.

$$9 = \frac{9}{1} \qquad 3 = \frac{3}{1} \qquad 6 = \frac{6}{1} \qquad 7 = \frac{7}{1}$$

To divide a whole number by a fraction: $10 \div \frac{2}{3}$

STEP 1	STEP 2	STEP 3
Write the whole number as a fraction.	Multiply by the reciprocal.	Simplify and multiply.

$$10 = \frac{10}{1} \qquad\qquad \frac{10}{1} \times \frac{3}{2} \qquad\qquad \frac{\overset{5}{\cancel{10}}}{1} \times \frac{3}{\underset{1}{\cancel{2}}} = \frac{15}{1} = 15$$

Divide the whole number by the fraction.

1. $9 \div \frac{1}{3} = \frac{9}{1} \times \frac{3}{1} =$

2. $8 \div \frac{2}{3} =$

3. $7 \div \frac{2}{5} =$

4. $15 \div \frac{5}{6} =$

5. $14 \div \frac{2}{3} =$

6. $4 \div \frac{1}{5} =$

7. $5 \div \frac{5}{6} = \frac{5}{1} \times \frac{6}{5} =$

8. $11 \div \frac{1}{2} =$

9. $9 \div \frac{3}{4} =$

10. $21 \div \frac{3}{7} =$

11. $6 \div \frac{2}{5} =$

12. $6 \div \frac{5}{6} =$

Divide by Whole Numbers

When you divide by a whole number, you must find its reciprocal.

$$5 = \frac{5}{1} \quad \text{The reciprocal of } \frac{5}{1} \text{ is } \frac{1}{5}$$

$$3 = \frac{3}{1} \quad \text{The reciprocal of } \frac{3}{1} \text{ is } \frac{1}{3}$$

$$1\frac{1}{2} \div 6 = \frac{3}{2} \div \frac{6}{1} = \frac{\cancel{3}}{2} \times \frac{1}{\cancel{6}} = \frac{1}{4}$$

reciprocal — multiply

Divide the mixed number by the whole number. Write the answers in the simplest form.

1. $2\frac{1}{3} \div 4 = \frac{7}{3} \times \frac{1}{4} =$

2. $2\frac{1}{4} \div 6 =$

3. $\frac{3}{5} \div 5 =$

4. $1\frac{1}{5} \div 3 =$

5. $8\frac{1}{3} \div 5 =$

6. $4\frac{2}{5} \div 2 =$

7. $4\frac{2}{7} \div 5 = \frac{\overset{6}{\cancel{30}}}{7} \times \frac{1}{\underset{1}{\cancel{5}}} =$

8. $1\frac{3}{4} \div 2 =$

9. $7\frac{1}{2} \div 3 =$

10. $1\frac{1}{9} \div 2 =$

11. $6\frac{2}{3} \div 5 =$

12. $3\frac{1}{2} \div 2 =$

Divide by Mixed Numbers

When dividing by a mixed number:

- change the mixed number to an improper fraction
- multiply by the reciprocal of the improper fraction

$$\frac{3}{4} \div 2\frac{1}{2} = \frac{3}{4} \div \frac{5}{2} = \frac{3}{4} \times \frac{2}{5} = \frac{3}{10}$$

Divide by the mixed numbers. Write the answers in the simplest form.

1. $\frac{3}{8} \div 1\frac{1}{2}$

   $\frac{3}{8} \div \frac{3}{2} = \frac{\cancel{3}}{\cancel{8}} \times \frac{\cancel{2}}{\cancel{3}} = \frac{1}{4}$

7. $\frac{5}{6} \div 1\frac{1}{4}$

   $\frac{5}{6} \div \frac{5}{4} = \frac{5}{6} \times \frac{4}{5} =$

2. $\frac{1}{2} \div 2\frac{1}{2}$

8. $\frac{2}{5} \div 2\frac{2}{5}$

3. $\frac{1}{5} \div 3\frac{1}{5}$

9. $\frac{3}{7} \div 1\frac{1}{6}$

4. $\frac{3}{5} \div 2\frac{1}{2}$

10. $\frac{3}{8} \div 4\frac{5}{8}$

5. $\frac{1}{6} \div 1\frac{1}{9}$

11. $\frac{2}{3} \div 2\frac{1}{4}$

6. $\frac{7}{8} \div 3\frac{3}{4}$

12. $\frac{1}{2} \div 2\frac{3}{4}$

Divide Two Mixed Numbers

When dividing a mixed number by a mixed number, change each mixed number to an improper fraction and then multiply by the reciprocal of the second improper fraction.

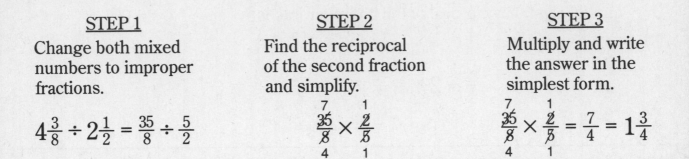

STEP 1
Change both mixed numbers to improper fractions.

$$4\frac{3}{8} \div 2\frac{1}{2} = \frac{35}{8} \div \frac{5}{2}$$

STEP 2
Find the reciprocal of the second fraction and simplify.

$$\overset{7}{\underset{4}{\cancel{\frac{35}{8}}}} \times \overset{1}{\underset{1}{\cancel{\frac{2}{5}}}}$$

STEP 3
Multiply and write the answer in the simplest form.

$$\overset{7}{\underset{4}{\cancel{\frac{35}{8}}}} \times \overset{1}{\underset{1}{\cancel{\frac{2}{5}}}} = \frac{7}{4} = 1\frac{3}{4}$$

Follow the steps above to solve each problem.

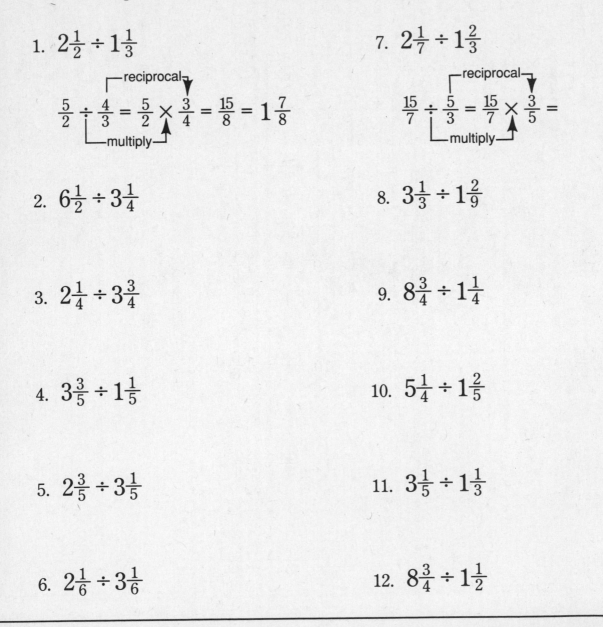

1. $2\frac{1}{2} \div 1\frac{1}{3}$

    reciprocal

   $\frac{5}{2} \div \frac{4}{3} = \frac{5}{2} \times \frac{3}{4} = \frac{15}{8} = 1\frac{7}{8}$

    multiply

2. $6\frac{1}{2} \div 3\frac{1}{4}$

3. $2\frac{1}{4} \div 3\frac{3}{4}$

4. $3\frac{3}{5} \div 1\frac{1}{5}$

5. $2\frac{3}{5} \div 3\frac{1}{5}$

6. $2\frac{1}{6} \div 3\frac{1}{6}$

7. $2\frac{1}{7} \div 1\frac{2}{3}$

    reciprocal

   $\frac{15}{7} \div \frac{5}{3} = \frac{15}{7} \times \frac{3}{5} =$

    multiply

8. $3\frac{1}{3} \div 1\frac{2}{9}$

9. $8\frac{3}{4} \div 1\frac{1}{4}$

10. $5\frac{1}{4} \div 1\frac{2}{5}$

11. $3\frac{1}{5} \div 1\frac{1}{3}$

12. $8\frac{3}{4} \div 1\frac{1}{2}$

Division Practice

To divide you must:

1. Change all whole numbers to fractions.
2. Change all mixed numbers to improper fractions.
3. Find the reciprocal of the second number and multiply.
4. Simplify whenever possible.

Examples:

$$6 \div \frac{1}{2} = \frac{6}{1} \times \frac{2}{1} = \frac{12}{1} = 12$$

$$4\frac{1}{2} \div 3 = \frac{9}{2} \div \frac{3}{1} = \frac{\overset{3}{\cancel{9}}}{2} \times \frac{1}{\underset{1}{\cancel{3}}} = \frac{3}{2} = 1\frac{1}{2}$$

Think carefully about these mixed division problems. Find the reciprocal of the second number and multiply. Be sure to simplify whenever possible.

1. $\frac{3}{5} \div \frac{1}{10}$

2. $3 \div 4\frac{2}{5}$

3. $\frac{2}{3} \div 2\frac{1}{4}$

4. $5\frac{1}{4} \div 1\frac{2}{5}$

5. $8\frac{1}{3} \div 3$

6. $7\frac{1}{2} \div 1\frac{1}{4}$

7. $\frac{3}{10} \div 3$

8. $6\frac{2}{3} \div 5$

9. $2\frac{4}{9} \div 1\frac{2}{3}$

10. $1\frac{7}{8} \div \frac{3}{4}$

Division Review

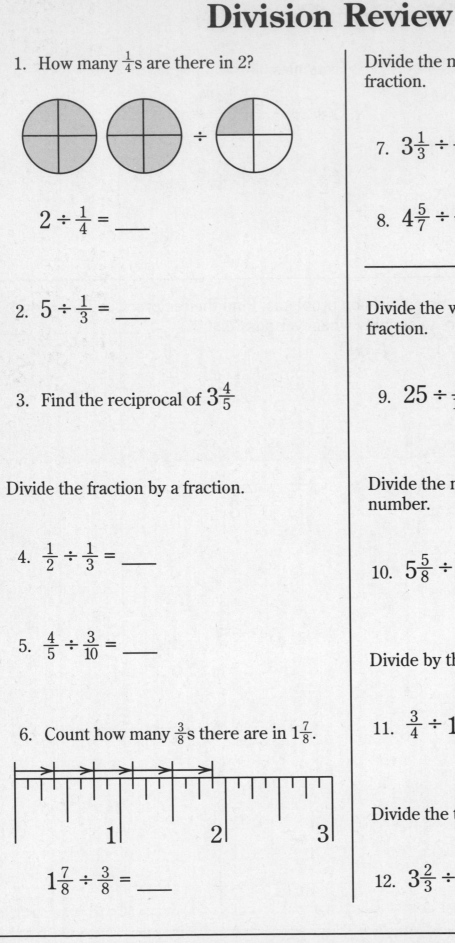

1. How many $\frac{1}{4}$s are there in 2?

$2 \div \frac{1}{4} =$ _____

2. $5 \div \frac{1}{3} =$ _____

3. Find the reciprocal of $3\frac{4}{5}$

Divide the fraction by a fraction.

4. $\frac{1}{2} \div \frac{1}{3} =$ _____

5. $\frac{4}{5} \div \frac{3}{10} =$ _____

6. Count how many $\frac{3}{8}$s there are in $1\frac{7}{8}$.

$1\frac{7}{8} \div \frac{3}{8} =$ _____

Divide the mixed number by the fraction.

7. $3\frac{1}{3} \div \frac{7}{9} =$ _____

8. $4\frac{5}{7} \div \frac{11}{14} =$ _____

Divide the whole number by the fraction.

9. $25 \div \frac{5}{11} =$ _____

Divide the mixed number by the whole number.

10. $5\frac{5}{8} \div 9 =$ _____

Divide by the mixed number.

11. $\frac{3}{4} \div 1\frac{1}{4} =$ _____

Divide the two mixed numbers.

12. $3\frac{2}{3} \div 1\frac{5}{6} =$ _____

42

Use All Operations

Use all of the operations with fractions. Write your answers in simplest form.

1. $3\frac{1}{2}$
    $-1\frac{1}{10}$

7. $4\frac{2}{3}$
    $+9\frac{3}{4}$

2. $\frac{9}{16}$
    $+\frac{1}{8}$

8. $8\frac{1}{6}$
    $-3\frac{5}{12}$

3. $4\frac{3}{4} \times \frac{4}{5} =$

9. $3\frac{1}{3} \div 3\frac{1}{8} =$

4. $\frac{1}{2} \div \frac{4}{5} =$

10. $9 \times 1\frac{2}{3} =$

5. $18\frac{3}{4}$
    $-11\frac{1}{12}$

11. $4\frac{5}{6}$
    $+3\frac{5}{12}$

6. $2\frac{2}{3} \times 3\frac{3}{8} =$

12. $\frac{2}{7} \div 1\frac{1}{3} =$

Putting It All Together

Place the symbols < , >, or = in the ◯ to make each statement true.

1. $\underbrace{\frac{1}{8} \div \frac{1}{2}}_{\frac{1}{4}}$ **(<)** $\underbrace{1\frac{1}{3} \div \frac{2}{5}}_{3\frac{1}{3}}$

2. $\frac{6}{7} \times 3\frac{8}{9}$ ◯ $8\frac{1}{2} \div 2\frac{3}{4}$

3. $10\frac{1}{9} - 4\frac{5}{6}$ ◯ $2\frac{4}{5} + 1\frac{9}{10}$

4. $5\frac{1}{4} \div 1\frac{1}{6}$ ◯ $8\frac{3}{4} \div 2\frac{7}{8}$

5. $4\frac{5}{8} + 1\frac{1}{4}$ ◯ $7\frac{3}{4} - 2\frac{3}{8}$

6. $\frac{5}{8} - \frac{1}{2}$ ◯ $\frac{1}{8} + \frac{1}{4}$

7. $4\frac{5}{8} + 1\frac{1}{4}$ ◯ $6\frac{3}{4} - 2\frac{3}{8}$

8. $\frac{3}{8} + \frac{7}{16}$ ◯ $6\frac{3}{4} \div 4\frac{1}{2}$

Does the Answer Make Sense?

1. Read over the problem several times to make sure you understand it.
2. Think about the facts in the problem and what you are being asked to find.
3. Complete the number sentence for each problem.
4. Ask yourself, "Does the answer make sense?"

---

1. Mr. Reinhold has a strip of wood 4 feet long. How many $\frac{1}{4}$-foot strips can be cut from the 4-foot strip?

$$\underline{\hspace{1cm}} \quad \underline{\hspace{0.6cm}}_{\substack{\text{operation} \\ \text{symbol}}} \quad \underline{\hspace{1cm}} \quad = \quad \underline{\hspace{1cm}}_{\text{answer}}$$

Mr. Reinhold can cut _____ strips from the 4-foot strip.

---

2. If $1\frac{1}{4}$ cups of sugar per gallon are called for in a $3\frac{3}{4}$ gallon punch recipe, how many cups of sugar are needed?

$$\underline{\hspace{1cm}} \quad \underline{\hspace{0.6cm}}_{\substack{\text{operation} \\ \text{symbol}}} \quad \underline{\hspace{1cm}} \quad = \quad \underline{\hspace{1cm}}_{\text{answer}}$$

_____ cups of sugar are needed for the punch recipe.

---

3. How many $\frac{3}{4}$ pound boxes of cookies can you get out of 12 pounds?

$$\underline{\hspace{1cm}} \quad \underline{\hspace{0.6cm}}_{\substack{\text{operation} \\ \text{symbol}}} \quad \underline{\hspace{1cm}} \quad = \quad \underline{\hspace{1cm}}_{\text{answer}}$$

You can get _____ boxes of cookies out of 12 pounds.

---

4. Sally must mix $1\frac{1}{4}$ quarts of oil for every gallon of gasoline. How many quarts of oil will she use for $7\frac{1}{2}$ gallons of gasoline?

$$\underline{\hspace{1cm}} \quad \underline{\hspace{0.6cm}}_{\substack{\text{operation} \\ \text{symbol}}} \quad \underline{\hspace{1cm}} \quad = \quad \underline{\hspace{1cm}}_{\text{answer}}$$

_____ quarts of oil will be used for $7\frac{1}{2}$ gallons of gasoline.

---

5. How many $\frac{3}{4}$-foot boards can be cut from 15 feet?

$$\underline{\hspace{1cm}} \quad \underline{\hspace{0.6cm}}_{\substack{\text{operation} \\ \text{symbol}}} \quad \underline{\hspace{1cm}} \quad = \quad \underline{\hspace{1cm}}_{\text{answer}}$$

You can cut _____ $\frac{3}{4}$-foot boards from 15 feet.

---

6. One centimeter is about $\frac{2}{5}$ of an inch. About how many centimeters are there in 6 inches?

$$\underline{\hspace{1cm}} \quad \underline{\hspace{0.6cm}}_{\substack{\text{operation} \\ \text{symbol}}} \quad \underline{\hspace{1cm}} \quad = \quad \underline{\hspace{1cm}}_{\text{answer}}$$

There are about _____ centimeters in 6 inches.

Decide to Multiply or Divide

It may be hard to decide whether to multiply or divide in a fraction problem. It helps to think about the following types of problems.

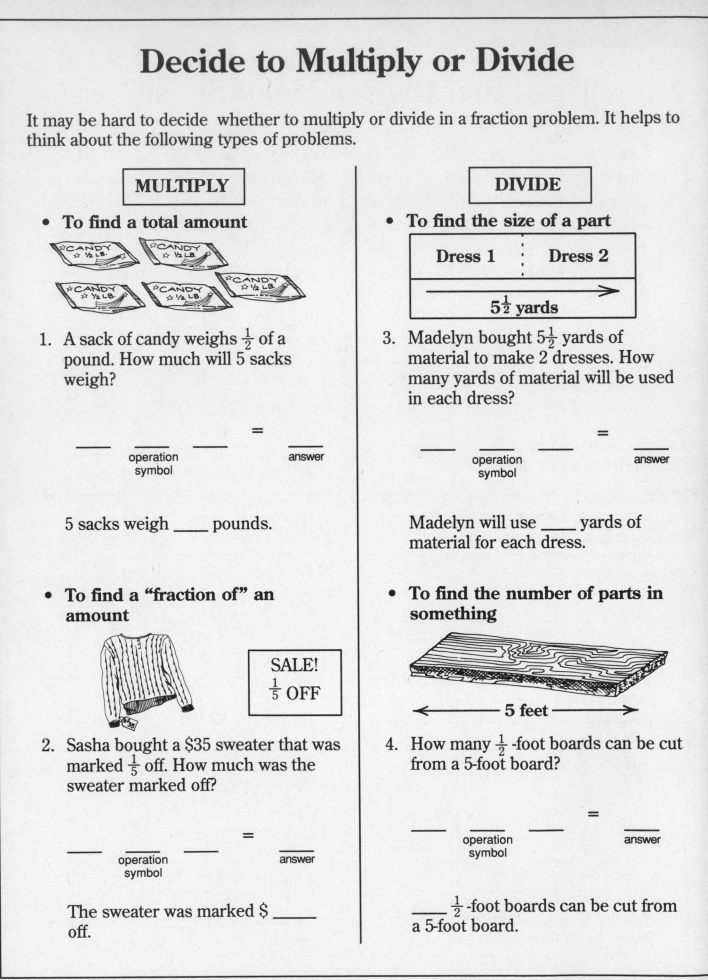

MULTIPLY	DIVIDE

MULTIPLY

- **To find a total amount**

1. A sack of candy weighs $\frac{1}{2}$ of a pound. How much will 5 sacks weigh?

   ___ ___ ___ = ___
   operation   answer
   symbol

   5 sacks weigh ___ pounds.

- **To find a "fraction of" an amount**

   SALE! $\frac{1}{5}$ OFF

2. Sasha bought a $35 sweater that was marked $\frac{1}{5}$ off. How much was the sweater marked off?

   ___ ___ ___ = ___
   operation   answer
   symbol

   The sweater was marked $ ___ off.

DIVIDE

- **To find the size of a part**

Dress 1	Dress 2

   $5\frac{1}{2}$ yards

3. Madelyn bought $5\frac{1}{2}$ yards of material to make 2 dresses. How many yards of material will be used in each dress?

   ___ ___ ___ = ___
   operation   answer
   symbol

   Madelyn will use ___ yards of material for each dress.

- **To find the number of parts in something**

   ←— 5 feet —→

4. How many $\frac{1}{2}$-foot boards can be cut from a 5-foot board?

   ___ ___ ___ = ___
   operation   answer
   symbol

   ___ $\frac{1}{2}$-foot boards can be cut from a 5-foot board.

Mixed Multiplication and Division

Read each problem carefully. Decide whether to multiply or divide. Write the answer in the simplest form.

1. Nate divided a $3\frac{1}{2}$-inch line into $\frac{1}{4}$-inch parts. How many parts were there?

   _____ _____ _____ = _____
   operation   answer
   symbol

   There were _____ parts.

2. A tank holds $7\frac{1}{2}$ gallons. If the tank is $\frac{1}{5}$ full, how many gallons are in the tank?

   _____ _____ _____ = _____
   operation   answer
   symbol

   There are _____ gallons left in the tank.

3. Alex walked $3\frac{1}{3}$ miles each hour for $2\frac{1}{4}$ hours. How many miles did he walk?

   _____ _____ _____ = _____
   operation   answer
   symbol

   Alex walked _____ miles in $2\frac{1}{4}$ hours.

4. Jodi used $1\frac{1}{2}$ pounds of flour to make 2 loaves of bread. How many pounds of flour did she use for each loaf?

   _____ _____ _____ = _____
   operation   answer
   symbol

   Jodi used _____ pound of flour for each loaf of bread.

5. How many rows $2\frac{1}{2}$ feet wide can be planted in a garden that is 10 feet wide?

   _____ _____ _____ = _____
   operation   answer
   symbol

   There will be _____ rows.

6. Silas worked five $8\frac{1}{2}$-hour shifts. How many hours did he work in all?

   _____ _____ _____ = _____
   operation   answer
   symbol

   Silas worked _____ hours in all.

7. Matt rode his bicycle 25 miles in $2\frac{1}{2}$ hours. How many miles did he travel in one hour?

   _____ _____ _____ = _____
   operation   answer
   symbol

   Matt rode his bicycle _____ miles in one hour.

8. A recipe calls for $\frac{3}{4}$ cup of cooking oil. Karen wants to double the recipe. How many cups of cooking oil does she need?

   _____ _____ _____ = _____
   operation   answer
   symbol

   Karen needs _____ cups of cooking oil.

Think It Through

To decide which operation to use, you must read carefully. One way to learn to read carefully is to write your own questions.

1. A recipe calls for $3\frac{1}{2}$ cups of flour and $1\frac{1}{4}$ cups of milk.

   Write a question about the facts if the answer is:

   a) $4\frac{3}{4}$ cups _____

   _____

   b) $2\frac{1}{4}$ cups _____

   _____

2. Lacey bought a $24.00 sweater that was marked $\frac{1}{4}$ off.

   Write a question about the facts if the answer is:

   a) $6.00 _____

   _____

   b) $18.00 _____

   _____

3. Bob worked an $8\frac{1}{2}$-hour shift.

   Write a question about the facts if the answer is:

   a) 17 hours _____

   _____

   b) $25\frac{1}{2}$ hours _____

   _____

4. Bonita traveled 108 miles in $2\frac{1}{4}$ hours.

   Write a question about the facts if the answer is:

   a) 48 miles _____

   _____

   b) 240 miles _____

   _____

5. One centimeter is about $\frac{2}{5}$ of an inch.

   Write a question about the facts if the answer is:

   a) 2 inches _____

   _____

   b) $1\frac{1}{5}$ inches _____

   _____

6. 3 boxes weigh $4\frac{1}{2}$ pounds.

   Write a question about the facts if the answer is:

   a) $1\frac{1}{2}$ pounds _____

   _____

   b) $7\frac{1}{2}$ pounds _____

   _____

Write a Question

1. Peggy bought a $36 sweater that was marked $\frac{1}{4}$ off.

   a) Question: _How much money was marked off?_

   b) ____ $\times$ ____ = ____
      <small>operation symbol</small>   <small>answer</small>

4. Each sack of candy weighs $\frac{1}{2}$ of a pound. Hilda has 12 sacks of candy.

   a) Question: _____

   _____

   b) ____ ____ = ____
      <small>operation symbol</small>   <small>answer</small>

2. Terry walked $1\frac{1}{2}$ miles each day for 14 days.

   a) Question: _____

   _____

   b) ____ ____ = ____
      <small>operation symbol</small>   <small>answer</small>

5. One centimeter is $\frac{2}{5}$ of an inch. A line segment measures 20 inches.

   a) Question: _____

   _____

   b) ____ ____ = ____
      <small>operation symbol</small>   <small>answer</small>

3. Mr. Links wants to cut a 6-foot-long piece of wood into $1\frac{1}{2}$-foot pieces.

   a) Question: _____

   _____

   b) ____ ____ = ____
      <small>operation symbol</small>   <small>answer</small>

6. Lynn saves $\frac{1}{5}$ of her $125 earnings each week.

   a) Question: _____

   _____

   b) ____ ____ = ____
      <small>operation symbol</small>   <small>answer</small>

Apply the Operations

Match the letter to the correct phrase.

_____   1. To find how many equal groups there
letter   are in a total   A. add

_____   2. To compare two numbers to find the   B. subtract
letter   difference in amounts
                                                         C. multiply
_____   3. To combine two or more different numbers
letter   to find the total of something   D. divide

_____   4. To increase or join together many
letter   like things or groups

Read each problem carefully and decide which operation to use. Write a number
sentence for each problem.

5. A nut bread recipe calls for $1\frac{1}{4}$ cups
   of chopped nuts. If Kerri wants to
   make $\frac{1}{2}$ of the recipe, how many
   cups of chopped nuts will she need?

   ___ ___ ___ = ___
   operation   answer
   symbol

   Kerri will need ____ cup of
   chopped nuts.

6. Mason walked 15 miles in $3\frac{3}{4}$ hours.
   At the same rate, how many miles
   can he walk in one hour?

   ___ ___ ___ = ___
   operation   answer
   symbol

   Mason can walk ____ miles in one
   hour.

7. Mr. Gove spent $1\frac{1}{2}$ hours studying
   and $3\frac{3}{4}$ hours playing golf. How
   much time has he spent studying
   and playing golf?

   ___ ___ ___ = ___
   operation   answer
   symbol

   Mr. Gove spent ____ hours
   studying and playing golf.

8. A carpenter cuts $1\frac{1}{8}$ inches from an
   $8\frac{1}{2}$-inch board. How long is the
   remaining piece?

   ___ ___ ___ = ___
   operation   answer
   symbol

   The remaining piece is ____ inches.

Choose the Operation

Indicate which operation—**addition, subtraction, multiplication,** or **division**—should be used to solve each of the following problems. **Do not solve the problem.**

_____
operation

1. The shop was open $8\frac{1}{2}$ hours on Monday and $10\frac{3}{4}$ hours on Tuesday. How many hours did the shop stay open?

_____
operation

2. A board is $6\frac{3}{4}$ feet in length. It is $1\frac{1}{2}$ feet too long. After cutting off $1\frac{1}{2}$ feet, how long will the board be?

_____
operation

3. Bob bought a $12 sweater that was marked $\frac{1}{2}$ off. By how much was the sweater marked off?

_____
operation

4. How much will $\frac{1}{2}$ pound of meat cost at $.98 per pound?

_____
operation

5. Laurie used $\frac{1}{2}$ pound of cheese to make 4 sandwiches. How much cheese did she use per sandwich?

_____
operation

6. Mr. Miller watched television $2\frac{1}{2}$ hours on Saturday and $1\frac{1}{4}$ hours on Sunday. How long did he watch television?

_____
operation

7. Mr. Makowski has a strip of wood 6 feet long. How many $\frac{1}{3}$-foot strips can be cut from the 6-foot strip?

_____
operation

8. Amy drinks $1\frac{1}{2}$ glasses of orange juice each morning for breakfast. How much does she drink in 5 days?

_____
operation

9. Trina ran the first mile in $7\frac{1}{4}$ minutes. She ran the second mile in $8\frac{1}{2}$ minutes. How much faster did she run the first mile?

_____
operation

10. What is the total weight of two items weighing $3\frac{1}{5}$ pounds and $7\frac{1}{3}$ pounds each?

Mixed Problem Solving

Read each problem carefully. Decide whether to add, subtract, multiply, or divide.

1. Sam had 8 rows of strawberries to pick. He has already picked 5 rows.

   a) What fraction of the strawberries has Sam picked? ____
   fraction

   b) What fraction of the strawberries does Sam have left? ____
   fraction

2. It took Lyle $3\frac{1}{2}$ hours to mow Mr. Brown's lawn and $1\frac{1}{4}$ hours to mow Mr. Jellnick's lawn. How much longer did it take Lyle to mow Mr. Brown's lawn?

   ____ ____ ____ = ____
   operation symbol   answer

   It took Lyle ____ hours longer to mow Mr. Brown's lawn.

3. Kendra bought a $3\frac{1}{4}$-pound roast and $2\frac{1}{3}$-pound ham. What was the total weight of her purchase?

   ____ ____ ____ = ____
   operation symbol   answer

   The total weight of her purchase was ____ pounds.

4. A cupcake recipe calls for $\frac{2}{3}$ cup of sugar. If you make $\frac{1}{2}$ of the recipe, how many cups of sugar will you use?

   ____ ____ ____ = ____
   operation symbol   answer

   You will use ____ cup of sugar.

5. The Williams family spends $\frac{1}{10}$ of its monthly income on recreation and $\frac{1}{5}$ on rent. What part of the monthly income was spent on recreation and rent?

   ____ ____ ____ = ____
   operation   answer

   ____ was spent on recreation and rent.

6. How many $\frac{1}{2}$-foot boards can be cut from a board 4 feet long?

   ____ ____ ____ = ____
   operation symbol   answer

   ____ boards, $\frac{1}{2}$ foot in length, can be cut from a board 4 feet long.

Throw Away Extra Information

To successfully solve word problems you must:

1. Examine the facts.
2. Throw away all information that is not needed.
3. Choose a course of action.
4. Ask yourself, "Does the answer make sense?"

Cross out and list the fact(s) that are not needed to work the problem.

1. Carrie bought $2\frac{1}{2}$ pounds of chocolate for $3.98. She used $1\frac{1}{4}$ pounds in a recipe that serves 7 people. How many pounds of chocolate does she have left?

   a) Facts not needed: _____

   b) Why? _____

   _____

   c) Carrie has ____ pounds of chocolate left.

2. Delbert traveled 225 miles in $4\frac{1}{3}$ hours on Monday and 485 miles in $8\frac{1}{2}$ hours on Tuesday. How many hours did he travel altogether?

   a) Facts not needed: _____

   b) Why? _____

   _____

   c) Delbert traveled ____ hours altogether.

3. Edna bought $5\frac{1}{2}$ pounds of cheese for $15.98. Each pound serves 4 people. She invited 12 friends to the party. After the party she had $1\frac{3}{4}$ pounds of cheese left. How many pounds of cheese did she use?

   a) Facts not needed: _____

   b) Why? _____

   _____

   c) Edna used ____ pounds of cheese at the party.

Two-Step Story Problems

Use the answer you find in the first question to help you answer the second question.

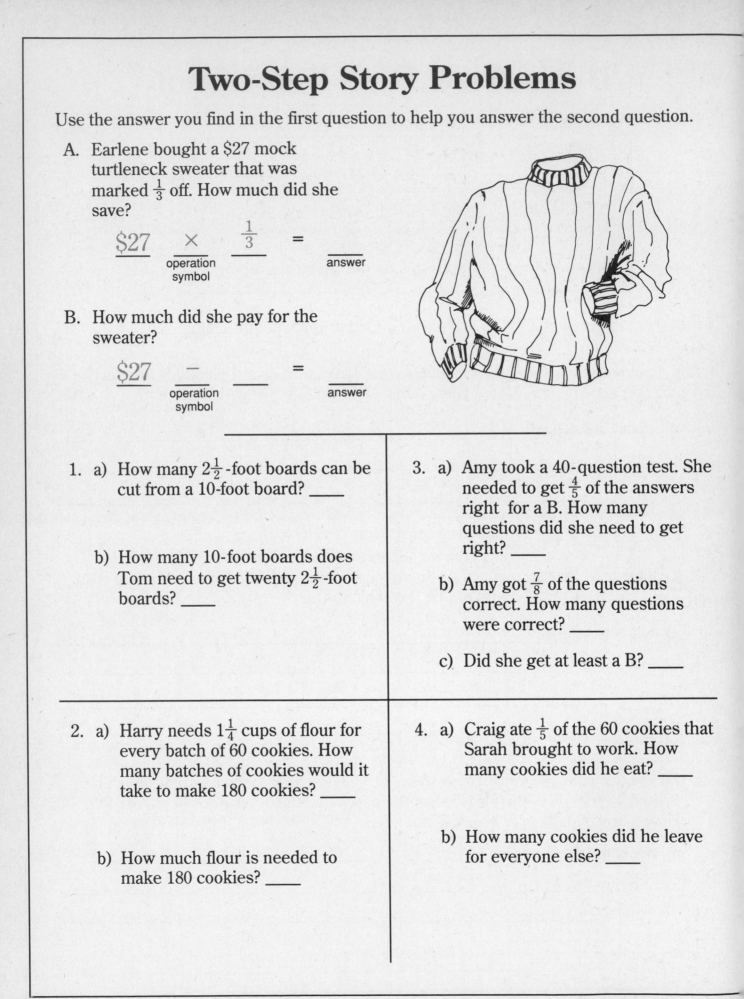

A. Earlene bought a $27 mock turtleneck sweater that was marked $\frac{1}{3}$ off. How much did she save?

$27 ____ $\times$ ____ $\frac{1}{3}$ = ____

operation symbol   answer

B. How much did she pay for the sweater?

$27 ____ $-$ ____ ____ = ____

operation symbol   answer

1. a) How many $2\frac{1}{2}$-foot boards can be cut from a 10-foot board? ____

   b) How many 10-foot boards does Tom need to get twenty $2\frac{1}{2}$-foot boards? ____

2. a) Harry needs $1\frac{1}{4}$ cups of flour for every batch of 60 cookies. How many batches of cookies would it take to make 180 cookies? ____

   b) How much flour is needed to make 180 cookies? ____

3. a) Amy took a 40-question test. She needed to get $\frac{4}{5}$ of the answers right for a B. How many questions did she need to get right? ____

   b) Amy got $\frac{7}{8}$ of the questions correct. How many questions were correct? ____

   c) Did she get at least a B? ____

4. a) Craig ate $\frac{1}{5}$ of the 60 cookies that Sarah brought to work. How many cookies did he eat? ____

   b) How many cookies did he leave for everyone else? ____

More Two-Step Problems

Do both steps to solve each problem.

A. Ivan bought 9 gallons of paint. If he used $2\frac{1}{2}$ gallons the first day and $4\frac{1}{4}$ gallons the next day, how many gallons did he use?

$\underset{\text{operation}\atop\text{symbol}}{\underline{\quad 2\frac{1}{2} \quad}} + \underline{\quad 4\frac{1}{4} \quad} = \underset{\text{answer}}{\underline{\qquad\qquad}}$

B. How many gallons are left? ____

$\underset{\text{operation}\atop\text{symbol}}{\underline{\quad 9 \quad}} - \underline{\qquad} = \underset{\text{answer}}{\underline{\qquad}}$

---

1. There are 32 students in Tina's math class. Out of $\frac{1}{4}$ of the students that are absent, $\frac{1}{2}$ have the flu. How many students in that class have the flu?

   ____ students in Tina's math class have the flu.

2. Janet takes home $250 a week. She always tries to set aside $\frac{1}{5}$ of her pay. Of the money she sets aside, Janet puts $\frac{1}{2}$ into her savings account. How much money does she add to her savings account each week?

   Janet adds ____ to her savings account each week.

3. Of the forty households in Martin's neighborhood, $\frac{3}{4}$ of them have pets. Of those pets, $\frac{2}{3}$ of them are dogs. How many households have dogs as pets?

   ____ of the households have dogs as pets.

4. A football team won $\frac{1}{2}$ of its 12 games. Of the games it won, $\frac{2}{3}$ were won by more than 14 points. How many games did the team win by more than 14 points?

   The team won ____ games by more than 14 points.

55

Multi-Step Word Problems

To solve problems with many steps, write several questions.

Nadine worked on her lessons $\frac{3}{4}$ of an hour on Monday and $1\frac{1}{2}$ hours on Tuesday. Jose' worked on his lessons $2\frac{1}{2}$ hours on Monday and $\frac{1}{4}$ of an hour on Tuesday. How many more hours did Jose' work than Nadine?

Question 1:   How many hours did Nadine work on her lessons? ____

Question 2:   How many hours did Jose' work on his lessons? ____

Question 3:   How many more hours did Jose' work than Nadine? ____

Write the questions and solve the problems on another sheet of paper.

1. Roberto has a weekly income of $435. If he spends $\frac{1}{3}$ of his income on rent and $\frac{1}{5}$ on food, how much money will he have left?

    Question 1: _____

    Question 2: _____

    Question 3: _____

3. Walter spent $8\frac{3}{4}$ hours painting the first day and $8\frac{1}{2}$ hours the second. It took Caren $9\frac{1}{4}$ hours the first day and $8\frac{3}{4}$ hours the second day to do the same amount of work. How much longer did Caren take to do her work?

    Question 1: _____

    Question 2: _____

    Question 3: _____

2. Three shifts make up a 24-hour day. If the first shift works $\frac{1}{4}$ of the day and the second shift works $\frac{1}{3}$ of the day, how many hours does the third shift work?

    Question 1: _____

    Question 2: _____

    Question 3: _____

4. Jaclyn bought $6\frac{1}{2}$ pounds of flour. She used $1\frac{1}{4}$ pounds for a bread recipe and $2\frac{3}{4}$ pounds for a cake recipe. How many times could Jaclyn fill the bread recipe with the flour that was left over?

    Question 1: _____

    Question 2: _____

    Question 3: _____

Decrease a Recipe

Honey Muffins

$1\frac{1}{2}$ cups flour $\frac{3}{4}$ cup milk

3 tsp. baking soda $\frac{1}{3}$ cup honey

$\frac{1}{4}$ cup sugar   2 eggs

$1\frac{1}{4}$ tsp. salt

1. Julie wants to decrease by $\frac{1}{2}$ the size of the recipe. How much of each ingredient should she use?

   a) _____ cup flour   e) _____ cup milk

   b) _____ tsp. baking soda   f) _____ cup honey

   c) _____ cup sugar   g) _____ egg

   d) _____ tsp. salt

2. One dozen muffins = 12

   If the original recipe makes 1 dozen muffins, how many muffins will $\frac{1}{2}$ the recipe make? _____

3. a) 1 dozen = _____

   b) With 1 dozen eggs, how many batches of honey muffins can Julie make? _____

Increase a Recipe

Chocolate Chip Cookies

$2\frac{1}{2}$ cups flour $\frac{1}{2}$ cup butter

$1\frac{1}{3}$ cups brown sugar   8 oz. chocolate

1 tsp. vanilla   chips

$\frac{1}{2}$ tsp. salt   2 eggs

$\frac{3}{4}$ tbsp. baking powder

1. Jack wants to double this recipe. How much of each ingredient will he need?

     a) _____ cups flour   f) _____ cup butter

     b) _____ cups brown sugar   g) _____ oz. chocolate chips

     c) _____ tsp. vanilla   h) _____ eggs

     d) _____ tsp. salt

     e) _____ tbsp. baking powder

2.   Jack has 5 cups of flour left. How many batches of cookies can he make? _____

FLOUR

5 lbs.

3. If Jack made 3 batches of his cookie recipe, how much of these ingredients would he need?

     a) _____ cups brown sugar

     b) _____ tsp. salt

At the Store

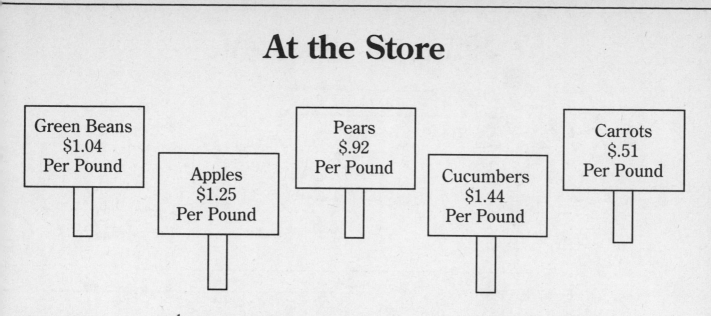

Green Beans
$1.04
Per Pound

Apples
$1.25
Per Pound

Pears
$.92
Per Pound

Cucumbers
$1.44
Per Pound

Carrots
$.51
Per Pound

To find the cost of $2\frac{1}{4}$ pounds of green beans at $1.04 per pound, you multiply.

$$2\frac{1}{4} \times 1.04 = \frac{9}{4} \times \frac{1.04}{1} = \frac{9 \times 1.04}{4} = \$2.34$$

Answer the questions using the prices shown in the picture.

Item	Cost Per Pound	Cost
$3\frac{1}{2}$ pounds of green beans	1. a) $1.04	b)
$2\frac{3}{4}$ pounds of pears	2. a)	b)
$1\frac{1}{3}$ pounds of carrots	3. a)	b)
7 pounds of apples	4. a)	b)
$1\frac{1}{8}$ pounds of cucumbers	5. a)	b)
	Total Cost	6.

7. Find the total cost of 2 pounds of green beans and 3 pounds of cucumbers. ____

8. How much more is $3\frac{1}{2}$ pounds of green beans than $1\frac{1}{8}$ pounds of cucumbers? ____

9. How much more is 4 pounds of apples than $1\frac{1}{2}$ pounds of pears? ____

10. What is the total cost of $\frac{1}{2}$ pound of green beans and $\frac{2}{3}$ pound of carrots? ____

Common Discounts

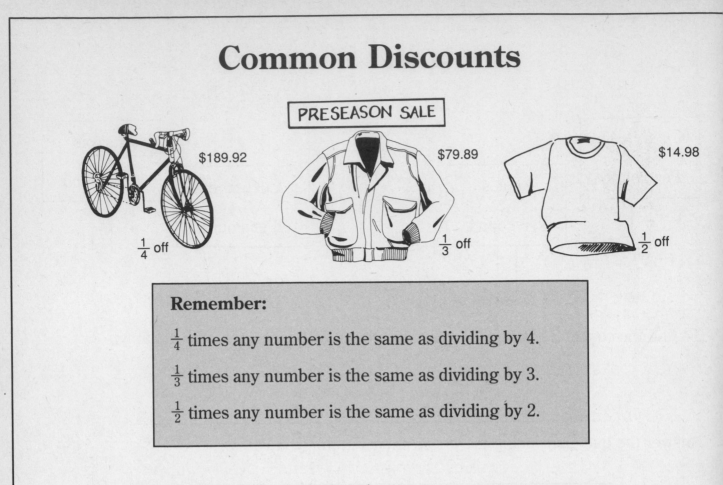

Remember:

$\frac{1}{4}$ times any number is the same as dividing by 4.

$\frac{1}{3}$ times any number is the same as dividing by 3.

$\frac{1}{2}$ times any number is the same as dividing by 2.

1. How much will the bicycle cost if it is $\frac{1}{4}$ off the regular price of $189.92? ____

2. How much can you save if the regular price of the jacket is $79.89 and it is marked $\frac{1}{3}$ off? ____

3. How much will the shirt cost if it is $\frac{1}{2}$ off the regular price of $14.98? ____

4. Find the discount on each full price.

	full price	$\frac{1}{4}$ off discount		full price	$\frac{1}{3}$ off discount		full price	$\frac{1}{2}$ off discount
a)	$32.00	_____	c)	$27.00	_____	e)	$98.00	_____
b)	$16.52	_____	d)	$3.99	_____	f)	$599.00	_____